D1805875

Acoustic Technics

Postphenomenology and the Philosophy of Technology
Series Editor-in-Chief: Robert Rosenberger, Georgia Institute of Technology
Executive Editors: Don Ihde, Stony Brook University, Emeritus and Peter-Paul Verbeek, University of Twente

Technological advances affect everything from our understandings of ethics, politics, and communication, to gender, science, and selfhood. Philosophical reflection on technology helps draw out and analyze the nature of these changes, and helps us understand both the broad patterns and the concrete details of technological effects. This book series provides a publication outlet for the field of the philosophy of technology in general, and the school of thought called "postphenomenology" in particular. Philosophy of technology applies insights from the history of philosophy to current issues in technology, and reflects on how technological developments change our understanding of philosophical issues. In response, postphenomenology analyzes human relationships with technologies, while integrating philosophical commitments of the American pragmatist tradition of thought.

Titles in the Series

Acoustic Technics

Don Ihde

LEXINGTON BOOKS
Lanham • Boulder • New York • London

Published by Lexington Books
An imprint of The Rowman & Littlefield Publishing Group, Inc.
4501 Forbes Boulevard, Suite 200, Lanham, Maryland 20706
www.rowman.com

Unit A, Whitacre Mews, 26-34 Stannary Street, London SE11 4AB

Copyright © 2016 by Lexington Books

All rights reserved. No part of this book may be reproduced in any form or by any
electronic or mechanical means, including information storage and retrieval systems,
without written permission from the publisher, except by a reviewer who may quote
passages in a review.

British Library Cataloguing in Publication Information Available

Library of Congress Cataloging-in-Publication Data

Names: Ihde, Don
Title: Acoustic technics / Don Ihde.
Description: Lanham,Maryland : Lexington Books, an imprint of The Rowman and Littlefield
    Publing Group, Inc, [2016]. | Includes bibliographical references and index.
Identifiers: LCCN 2015037020| ISBN 9781498519236 (cloth : alk. paper) | ISBN 9781498519243
    (electronic)
Subjects: LCSH: Acoustical engineering. | Ultrasonics. | Sound--Recording and reproducing.
Classification: LCC TA365 .I39 2015 | DDC 620.2--dc23 LC record available at http://lccn.loc.gov/
    2015037020

™ The paper used in this publication meets the minimum requirements of American
National Standard for Information Sciences Permanence of Paper for Printed Library
Materials, ANSI/NISO Z39.48-1992.

Printed in the United States of America

Dedicated to my animal friends in Vermont, as seen and heard through
my study window:
Acrobat Swallows
Grinning Otters
Diving Kingfishers
Elegant Blue Heron
Leaping Rainbow Trout
Lazy Swimming Turtles
Giant Moose Antlers Floating
Curious Black Bear
Frustrated Goshawk
Five Golden Eagles
Hypnotized Bullfrogs
Charmed Chickadees
Underwater Salamanders and Crayfish
And myriad others

Contents

List of Figures

Acknowledgments

Beginning from afar, I wish to acknowledge folk who have read parts of this book in preparation, only some of whom I have met while others remain online acquaintances. These include Karin Aras from Sweden, Nicola Miller from Scotland, and fellow auditory researchers Finn Olesen and Jette Claussen from Denmark, and Stacey Irwin from the United States. Add Barry Truax, Canada, and Inge Siegumfeld, Denmark, hosts for conferences on sound. Then, close to the project, Jana Hodges-Kluck and Robert Rosenberger, editors, have had many helpful suggestions. Robert also helped with word processing technical tasks. From earlier days while still at Stony Brook, Judith Lochhead, Robert Crease, and Eduardo Mendieta, colleagues, have helped my understanding of music and instrumentation, and to which I must add Trevor Pinch from Cornell, who has collaborated and written concerning so much in our common interests. The list could go on, but thanks, too, to those who contributed to Evan Selinger's *Postphenomenology: A Critical Companion to Ihde* (2006), to Lochhead and Pinch, add Lenore Langsdorf, former student, a long-time friend and interlocutor. Linda, my wife, sometimes techie aide, and our often mentioned, departed son, Mark, play strong roles, too.

Preface: Prereflections on science, technologies, visualization, and sonification

In 1976, I published *Listening and Voice: A Phenomenology of Sound.* This was my first original systematic foray into *doing* phenomenology and since it was also a reflection upon how we experience our world, I had to take account of what had been a dominantly *visualist* orientation in both philosophy and science in Western thought. This issue has returned repeatedly in the now more than forty years of reflections upon sight and sound in human experience and practice. *Acoustic Technics* is a very contemporary follow-up to these decades of reflections and returns to many of the same issues raised earlier, now with a closer look at hi-tech acoustic imaging technologies.

There is more to this history of reflective publication. As readers will know, *Technics and Praxis: A Philosophy of Technology* (1979) marked my turn to technologies, but as related to science and what is today more often called *technoscience.* In that book, I claimed that all science knowledge production is embodied in science's technologies, instruments. Early modern science was especially fond of *optical technologies,* telescopes and microscopes and variants upon the *camera obscura.* Following the histories of science and its instrumentation, it became obvious that this earlier dominance of *visualism* retained its hold upon science's knowledge production through vast stretches of time. In *Expanding Hermeneutics: Visualism in Science* (1998), I showed how science had over its centuries of development invented a very sophisticated *visual hermeneutics.* That is to say, a deep visualist interpretation of scientific phenomena through the use of dominantly visual imaging technologies. But images call for *interpretation* and interpretation is a *hermeneutics,* but not necessarily a linguistic or textual interpretation. Thus, I called for "expanding hermeneutics" into its visualist variations and showed how science had—in practice—developed such a hermeneutics.

One error I wanted to avoid from the beginning, was to switch from ocularcentrism to otocentrism. This for two very different reasons: first, it is obvious that science does not exclusively rely upon visualistic instrumental embodiments. In the history of medical instrumentation, for example, as Bettyann Kevles pointed out in her *Naked to the Bone* (1997) osculatory or listening instruments in medicine actually dominated much diagnostic practice: listening tubes, and other variants, eventually re-sulted in the stethoscope which is still iconic today. The highly symbolic stethoscope dangles from all TV doctor's necks just as it does from my own primary care doc. In practice it was used to detect everything from fetal movement to heart murmurs. Today there are articles bemoaning the loss of auditory skills to recognize such subtle acoustic images on the part of younger generations who are not taught these listening skills—instead they are more likely to take courses in radiology. Later, in the early twentieth century one major instrument developed during World War II, was sonar, an echo-location sounding device which detected ships and particularly submarines through the water. At first skilled us-ers of these devices learned to interpret acoustic signals (*pings* and *clicks*) before they were visualized to parallel radar, the above ground analog technology for aircraft. Sonar and radar today both image upon screens. Both these technologies, with superior technological development by the Allies, were crucial to winning the war.

The deeper reason to avoid a parallel reductionism to auditory phe-nomena compared to visual phenomena, lies in a phenomenological no-tion of the body through embodiment skills. I shall put this simply and starkly upfront here, while noting that such a sense of embodiment actu-ally is hard won from rigorous phenomenological—more, postpheno-menological analyses, and not a presupposed *apriori*—I would claim as a postphenomenologist, that all our perceptions are *whole body perceptions.* I made this point in *Listening and Voice,* but also in a chapter on rock music in *Technics and Praxis.*

Yes, we certainly hear with our ears—but not with our ears alone. Our ears are focal organs and they detect much of the crucial range of sound frequencies which are of human interest—but, for example we "self-hear" our own speech differently than others due to bone conduction. Our speaking vibrates our bone structure and thus we always sound "deeper" to ourselves than to others hearing us. And as any rock music listener knows, we hear the lower frequencies of bass notes "with our stomachs" and lower bodies as well as with our ears. Physiologically speaking, the very cavities inside our lower bodies become resonating chambers for these low tones which are "felt" as much as "heard." We sense with our whole bodies. This phenomenon will return as more im-portant than ever once we enter the realms of beyond body phenomena which science has opened up only since the nineteenth century. But I cannot resist a very contemporary example—in this case the discovery of whole body perception by contemporary sciences which converge with postphenomenological analyses.

For the decades in which I have done technoscience research I have been a devoted reader of those state-of-the-art science publications which produce reports on the latest research, including *Science, Scientific American* (for most of these years also *Nature,* given up only later on retirement), along with others less rigorously science-oriented such as *National Geographic, The Smithsonian, The Economist,* etc. Then, also in composing, fact checking online revealed how thoroughly and deeply science publication dominates much of the Internet with its particular Western master narrative perspective. Indeed, almost all the important research updates appear on the Internet in abbreviated and sometimes simplified forms.

It is from these sources, in very recent years, that there has appeared a growing awareness from physiology, neuro-physiology, that at least the neurological sense cells are not exclusively centered on our focal organs. One particular trick now used to identify such cell structures is the bio-tech ability to make such entitities "glow" in green fluorescent colors. The result has been the discovery that taste and smell buds show up in many unexpected locations! In an online article, "Full-body Taste" Science News for Students,[1] green glowing taste buds were discovered in the nose—but later in the gut, in the stomach, and even lungs. Indeed, the conclusion is, "there are more taste cells outside the mouth than inside the mouth."[2]

Postphenomenologically this should not be too surprising—anyone who has a connoisseur wine tasting capacity knows that we can "taste" with our noses and identify vintages, locales, and the particular tastes of a specific wine. I shall follow this convergence tactic repeatedly later. Here note that in spite of what I am calling a convergence, the style of biologically based neuroscience differs in strategy from that of postphenomenology. To claim "full-body taste," biology has to locate specific, localized, and narrow functioned entities as its evidence for experience. Postphenomenology analyzes perceptual experience to locate its possibility-range as a first beginning. It may, or may not be, that whole body perception is cell-dependent. Hearing rock bass notes may be as much felt as heard, and this ambiguity will show up again with sounds beyond sound.

So, now to the strategy of this book, *Acoustic Technics.* My move to a primarily auditory-acoustic focus will take place in stages. I begin in part 1 with a broad sweep which I call a Sono-Sightscape which includes both optical and acoustic technics. Although I have mentioned beginnings in the nineteenth century before, in this book I take a closer look at technoscience developments. Chapter 1 will actually focus upon the first scientifically noted awareness that there are sounds beyond sound and light beyond light and thus I open with the first experiments which evidenced infrared and ultraviolet in color emissions at the very opening of the nineteenth century. By mid-century what would become the framework for a whole range of radiation of which humans were unaware, the electromagnetic spectrum (EMS) began to open and by its end, x-rays, radio waves, and other now familiar emissions had been discovered.

Interestingly, sound beyond sound origins were much more ancient and embedded in Asian meditation practices.

If part 1 of *Acoustic Technics* is a sonoscape and sightscaple, or a landscape which locates areas for acoustic technologies (and other related technologies), it must work with the dominant visualism of science. Yet there are often surprising auditory moments. For example, in an editorial in *Science* concerning the interrelation between parks and science, the editor says, "The NPS recently used 1.5 million hours of acoustic data to create a soundscape map of the United States than can help guide the conservation of natural sounds."[3] In this book, and for the first time in my own narratives, I begin with the nineteenth century in this survey mode.

The nineteenth century was a big century for technology—it completed the first Industrial Revolution, contained the first modern origins of philosophy of technology, and formed, I would say, the outlines of a *triumphal* technoscience, including an interpretation of science as the highest form of knowledge ever reached by humankind. Chapter 2 continues the trajectory which filled out the electromagnetic spectrum of radiation and illustrates some of the instruments which could image this radiation all along the continuum of the spectrum from the longest radio to the shortest gamma waves—a "second scientific revolution" embodied in imaging technologies.

In chapter 3, I introduce a new suggestion for postphenomenology—I argue that we should vary not only technological variations, but include variations from animal studies—thus animals and robots. This allows a look at not only technological detectors which can mediate experiences other than our ordinary experience, but animal perceptions which either we lack entirely or have only very poorly. I introduce biomimetics as well, the adaptation of engineering design based upon animal bodily capacities, thus indirectly showing how embodiment issues pervade this field. Chapter 4 then focuses upon one special kind of experience, echolocation, which I hold we do experience and which we can develop and train, and contrast this with a long dominant, analytic approach to such border phenomena. In passing, I take note of some of the major notions of postphenomenology.

Following the broad scope of part 1, in part 2 I turn to a practice from the various science studies disciplines and STS or science-technology-studies approaches. These are "case studies" or dimensional studies which focus upon specific themes and technologies. I begin with a practice which likely goes to human origins: musics. This chapter, 5, is a deep history of musics in relation to changes in instrumentation with a post-phenomenological concern for variations. This chapter is reprinted from a feature article in *Janus Head* and contains a number of passing references found in other chapters, but since it is a crucial overall history, I did not try to eliminate these, but hope they will be accumulative. While I include many of my favorite "aha" moments in technoscience history, here I do so with a new context—musics. I follow this up, in chapter 6, with a narrower and newer set of cases—the development of synthesized sounds, often, but not exclusively musical. With chapter 7, I turn to medical acoustics, in this case the embodiment of hearing devices, followed by chapter 8 and a new acoustic technique of imaging, listening to cancer. I also inquire into the different ways in which art and science praxes related differently to their instruments. In chapter 9, "Acoustics below the Surface," I return to another continuum of sounds which range from maxi- to micro-waves related to subsurface phenomena, including a glimpse at surveillance techniques primarily in acoustic forms. Chapter 10, Multimedia-Multitasking-Multistability, will serve as a sort of summary about today's technology trajectories. I begin with two of the world's most ubiquitous technologies—television and cell phones—which are so common and widespread that they become almost invisible, but yet so powerful. Both are multimedia in the sense that they are what in media language are "audiovisual." After some posthphenomenology of these technologies, I shall return to very contemporary versions of high technologies in art, music, medicine, science, which today are tending to be opto-acoustic, or again, multimedia in display. Although short of what I would call whole body perceptibility, AV or opto-acoustic technologies are more than mono-dimensional. My claim will be that this is a technological trajectory which will change our ways of mediated experience of our lifeworlds. The text ends with an epilogue which raises the question of whether we are *posthuman*. I have appended, in highly abbreviated form, an appendix on technoscience in the twenty-first century which was part of a charter-founders panel at the 2013 meeting of the Society for Philosophy and Technology in Lisbon, Portugal. We were asked to reflect on the future of technological development, always a hazardous undertaking, and my appendix includes reflections relevant to this volume.

Finally, while previous readers will be familiar with my narrative style, for new readers note that unlike many academicians, I write in a first person style. My very early writings were more likely to be in a third person or anonymous voice, but I have been persuaded—mostly by feminist writers, including Susan Bordo, one of our alumna, and by Donna Haraway, so long a friend, that if one is serious about situated knowledge, especially one which forefronts embodiment as postphenomenology does, then a first person narrative is to be preferred. Part of my style has evolved with so many years and so much travel. This is clearly a "late life" book and hopefully will be read as an opening to conversation. Today's networking is often an alternation between electronically mediated communication and face-to-face encounters. Yet, so much of electronic communication has taken on highly informal and quickly familiar tone. First names are used in first contacts; anecdotes flow quickly; and while there are cultural differences—for example some cultures tend to take e-mail to be more "letter-like" while others take it as more "phone-like," exchanges are or can be rapid and frequent. All this has had much impact on how contemporary interconnected styles have emerged. Here, it is a book, but in years to come it could be a conversation.

NOTES

1. https://studentsocietyforsociety.org, "Full Body Taste."
2. *Ibid.*
3. Gary Maclis and Marcia McNutt, "Parks for Science," *Science,* Vo. 338, 19 June 2015, p. 1291

A Sono-Sightscape—Locating Acoustic Technologies

ONE

Sound beyond Sound and Light beyond Light

The narrative for this first chapter will begin with the year 1800 and end with the first Nobel Prize Award in 1901. In terms of technoscience history, the nineteenth century is the opening to the discovery of what today is known as the *electromagnetic spectrum*. This is, a vast spectrum of what we call *radiation* but which also is about the *emissions* which come to us from various sources and from different sectors of the entire universe. But experientially speaking this is also the discovery of radiation which goes beyond the limits of our experiential horizons—thus, sound beyond sound and light beyond light.

As I construct this narrative one major set of players will be those of the Western scientific canon—the discovers of previously unknown phenomena which set off quests for ever more like discoveries in the form of trajectories. But these players in the Western master narrative are not the only players. My history of science cannot, in the twenty-first century, retain its eurocentrism and must include discoveries made by others in other times and cultures as well. I have used sound and light to begin, but only because these are our familiar reference experiences.

The first scene opens with William Herschel's discovery of *infrared radiation*. As with so many discoveries, this one begins accidentally. Herschel and his sister, Caroline, were renowned astronomers in their own rights (Herschel discovered Uranus and Caroline was expert at detecting double stars, comets, and nebulae). Herschel was tinkering to see if he could produce some shading device to limit the heat which came with light through his then powerful instrument. He became interested in this phenomenon: light conveys heat and he began also to use a prism to test this association. He did notice—accidentally—that he *felt* heat even past the imaged red spectrum. But, in true scientific fashion for the time (and

3

before and past for that matter) he decided to *measure* this heat-light phenomenon. So he installed mercury thermometers with blackened bulbs along the spectrum lines. He discovered that the violet limit was the coldest; and red the hottest—but then astonishingly—heat emitted *beyond* the visible red limit was even hotter. *This was the first Western evidence for heat-light*—which he named "caloric rays"—beyond human visual perception. ("Ray" language was to dominate this century's optical discoveries.) Postphenomenologically I do not forget that he *felt* the heat although he did not *see* what we now call the infrared light. Whole body perception exceeds focal perception.

This discovery clearly lies centered in science experimentation. I remark that if one examines the history of astronomy, at least since early modernity (Galileo on) astronomy is marked by its embeddedness in optical instrumentation and human visual perception—and that its observations were limited to the optical spectrum—and to the distinct measurement practices which mark science's *measuring perceptions.* These, by the way, extend in this case to the thermometers which calibrate the heat of Herschel's "caloric rays." In my own multicultural histories of astronomy, I note that instrument-recorded quantities and instrumental standardizations well precede early modernity since solstices, calendars, eclipses, moon and sun cycles go back to the Ice Age at the least. But all astronomy until Herschel was optical or human light-perceiving astronomy.[1]

I cannot refrain from taking Herschel as a variant upon an earlier parallel discovery made in such a very similar fashion by Isaac Newton. Newton, in experiments in 1667 (a scant century earlier than Herschel), "I procured me a triangular glass prism to try the celebrated phenomenon of colors . . ." (with a hole in his "window shuts" and the image cast on the opposite wall, a variant upon the *camera obscura.*[2]) From this experiment Newton developed his theory of color—white light, composed of different, we would say wave frequencies, into the visible color spectrum of Newton, then on to the heat spectrum of Herschel here. Herschel finds that not only are the colors different, but the heat carrying capacities differ. Looking backward we can see that Herschel's discovery of light beyond (visible) light is potentially highly significant, although he clearly did not know what that significance was to become. He did clearly recognize that there was light beyond light. And he clearly had followed what Peter Galison calls an "instrumental tradition."[3]

In science, news travels fast. Owen Gingerich has published much about how fast Copernicus's heliocentric theory of celestial motion was distributed in the 1500s in his masterful book, *The Book Nobody Read* (2005). By the nineteenth century, science news was even faster. Johann Ritter had already heard of Herschel's discovery of infrared and theorized that there might well be a similar light beyond light at the blue-violet end of the spectrum. His 1801 experiments were radically different than

Herschel's; Ritter did not take heat as a prime variable. In retrospect, Ritter's variable was to parallel later photographic development. Ritter was a chemist-pharmacist and dabbled in light sensitive chemistry. Silver chloride was a chemical which turned black when exposed to sunlight and he was familiar with the fact that it responded more strongly to blue, than red light. (Note that this intensity scale is just the opposite to the temperature scale differences Herschel had noted.) Following now the more than century-old practice of using a prism to separate color frequencies, Ritter scattered silver chloride along the visible spectrum field and demonstrated that the farther into the blue, then violet, the black response intensified. So, now following the beyond-red trail of Herschel, Ritter placed silver chloride in the area beyond violet *and immediately discovered that well beyond its limits, the chemical showed intense reactions* (turning black). So, again following Herschel, he named these "chemical rays," later to become ultraviolet light. So, now from both ends of the light spectrum there is light beyond light, but also light which exceeds human, bodily perception.

This initial set of discoveries, 1800 and 1801, at the dawn of the nineteenth century brings us to one of the thematic problems for this book—how can phenomenology, better postphenomenology, deal with the newly discovered phenomena of light (soon also sound) beyond even our whole body capacities? My answer will come from the earliest move from classical to postphenomenology, the incorporation of "materiality," more specifically the embodied experience of technologies-instruments into intentionality itself. But first allow a brief pre-reflection on the narrative style I am using here: it will be noted that by paralleling Newton (late 1600s), Herschel and Ritter (early 1800s) that there is a long and deeply established instrumental praxis. Each used a prism to produce the focal phenomenon for the experiment, while drawing different implications from that phenomenon. Mainstream philosophy of science has largely ignored this pattern, although I tried to address it in my *Instrumental Realism* (1991) by showing how a few mainstream philosophers like Robert Ackerman, Ian Hacking, and Peter Galison do attend to instrumental praxis.

Physicists and astronomers will admit that much of the history of these sciences is bound up with light, optics, optical technologies. A former colleague of mine from Stony Brook, Arnold Goldhaber, even taught a course on the optical introduction to physics. But at the same time, this emphasis also reflects the ocularcentrism of so much science praxis—it is time to turn to the auditory dimension, sound. I have already remarked that the sciences do occasionally develop acoustic technologies which fit into their praxes of measuring perceptions—but these are not usually related to physics and astronomy which have played such a long dominant role as favored sciences. In this history there will be a major exception to be noted—radio-astronomy—but this does not take place in the nineteenth century which occupies this chapter. *Medical technologies* do occur. The acoustic technologies which culminate in the stethoscope are part of this history. And what such instruments present are part of the ordinary phenomenology of audition. Sounds reveal *interiors*, not only surfaces. Carpenters can tap walls to determine where the studs are placed behind the drywall; engine mechanics can use special versions of the stethoscope to detect valve-sounds of a running engine; and, as previously noted, the stethoscope can reveal heart murmurs and other internal sounds indicating bodily pathologies. Thus, there could be—was and is—an auditory hermeneutics which, with training and skill, becomes a diagnostic practice. But what about sound beyond sound?

The narrative context here has been the nineteenth century and that centered upon Western science history examples. These will return, but for sound the earliest discoveries of sound beyond sound of which I am aware come neither from science history, nor from Western history, but belong to Asian traditions and traditions which are steeped in religious, meditative, and auditory practices. This poses a problem for the narrative style I have been using—these practices are not chronologically or historically calibrated—thus, I cannot even give a date, nor an individual discoverer's name to origins. But it is well established that as part of Buddhist meditative (and there are analogs in Hindu practices) that ringing of gongs is a well-established part of meditative ritual at least a thousand years old. The sounds, reverberations, fill the meditator's listening and help to center a shaped concentration. But as the sounds continue, they begin to fade—until finally at some indistinct moment they can *no longer be heard*. Yet, *they can be felt* and participants have been encouraged to touch the gong and *feel* the continuing vibrations, now below the level of audible frequencies as we would say. As noted, I have been unable to find a person or date, but the practice is centuries old. I, myself, had this experience in a mountainside Buddhist temple in Kyoto, Japan, some years ago. Here, surely, is *sound beyond sound* and again related to the more complete whole body hearing which goes beyond what we can hear with our ears. Note that to feel the unheard sound vibrations, just like Herschel's felt heat, comes from beyond the focal sensory organs within experience.

So, it is also clearly analogous to the light beyond light phenomena noted in my previous narrative. Of course my experience did not relate to meditative practice, but to the bodily experience of sound. Thus, by the dawn of the nineteenth century, light and sound are found to exceed the human perceptual capacity to experience these "rays" and with a different language in sound with "waves."

But, phenomenologically speaking, this is also a major *phenomenological discovery*. Intentionality, which is the inter-relational framework of investigation used by both classical and postphenomenologists, takes account that for every change in a "world" there is a corresponding change in the experiencer, here the embodied human perceiver. In the case of sound beyond sound and light beyond light, this is the demonstration of *bodily perceptual horizons.* If we are indeed, situated, embodied beings here associating postphenomenology with the varieties of postmodernism which denies any ideal observer, "god's eye" or transcendent perspective, then the demonstration of horizons from within the perceivable is an equally important discovery. But it is a discovery *mediated by* precisely the instruments, the material techniques which show the limits. From first person experience horizons are also detectable as border limits. Thus, as I have shown in *Listening and Voice,* the border limit of a visual horizon is easily detectable simply by moving some object within the field of vision beyond its boundaries at the edges. And auditorily this is more difficult since one can hear omnidirectionally—the auditory field surrounds one. And thus, a horizon beyond which one can no longer hear is spatially a distance phenomenon, a sound is perceived to fade away with distance; although top and bottom frequency levels are also demonstrable—precisely through the use of instrumentation as illustrated in chapter 7.

For tactical reasons, I now want to jump to the very end of the nineteenth century with its shocking discovery of x-rays, which unlike the opening to infrared and ultraviolet which were both close to and analogous with experienced light. X-rays were not analogous to human perceptual experience, nor had they been humanly perceived. (Although some thought the strange green glows might be *ectoplasm*, the ghostly stuff of Victorian séance interests.) The actual discovery process was both more mundane and typical of the experimental tinkering of the nineteenth century. I again refer to Bettyann Kevles *Naked to the Bone* (1997), a masterful history of medical imaging. In the case of x-rays, instruments play a central role—William Crookes had invented the cathode ray tube in 1876 and this vacuum tube emitted "rays," still the language of the century. Conrad Roentgen, an experimentalist at the Polytechnic in Zurich, began a long, playful set of experiments with the green rays emitted by cathode ray tubes (Kevles points out that he was colorblind, thus did not recognize the rays as green).[4] Roentgen was also an inveterate visualist and liked to keep records—in this case photographic—of his discoveries, so he photographed results. X-rays thus, were produced by a compound technology: Crookes or cathode ray tubes plus photography. In his tinkering, Roentgen quickly recognized that the rays travelled distances; that they *penetrated* some materials, but not others; could have their images photographed and stabilized. The range of things with degrees of penetratability, he recognized, on photographic film were "shadows" and while metallic lead remained impenetrable, metal coins mostly so, but in yet another science "aha" moment, he discovered that his own bones also cast a shadow, as Kevles describes it, "Then he held his own hand to the invisible light—and became the first person in the world to see the shadow of his living bones."[5] But later, he took an x-ray of his wife's hand and her opaque ring, which made the photograph into post-cards which became the way he announced his discovery to the major scientists of the time.[6] The news was so astonishing and the response so quick that inventors began to produce x-ray machines all over the indus-trial world, uses of which spread from medicine to entertainment to the shoe stores many of us remember from our youth!

X-rays, though, were very different from the infrared-ultraviolet "rays" of the beginning of the nineteenth century. Infrared and ultravio-let were both understood to be "natural" radiations—usually from using sunlight—and were understood to be on an optical continuum from light. These ordinary light "rays" and the example of sound beyond sound carried over to felt experiences, beyond focal organs to be sure, but recog-nizable as belonging to experienceable phenomena. For x-rays, there was no such analogy (in spite of the ectoplasm speculation). No one had previously experienced x-rays. Today, if we use a sort of chronological "cheat code" of much later scientific knowledge, we know that there are natural sources of x-rays coming from astronomical sources—*but most of this radiation is blocked by earth's atmosphere which is opaque to x-rays, and, moreover, the sources are astronomical phenomena which were not known to early modern science: black holes, pulsars, supernovas, and their leftover nebu-lae.* And although a few of these have historically been visible (the Crab Nebula and other supernovae) no one experienced x–rays. This very short frequency particle-wave phenomenon lies beyond human percep-tual capacity—*except through instrumental mediation!*

And, that, too, was precisely a unique characteristic of these "rays" produced by Roentgen and his technologies. The green glow was pro-duced when current was turned on for the Crookes or cathode ray tube. One could say the phenomenon was produced *by* the apparatus and in any case it became experienceable only in instrumentally mediated form. Today, of course, x-rays belong to the vast continuum of the electromag-netic spectrum together with other ultra-short frequency "rays" such as gamma waves, an even later discovery. I have now begun with the dawn of the nineteenth century, 1800–1801, and the discoveries of infrared and ultraviolet, and skipped to its end, 1895, and the discovery of x-rays—which were awarded the very first Nobel Prize in 1901, the first of a whole series of Nobels for light and optical discoveries to the very present. But this chapter cannot end here since I have left its equally rich discovery center years unmentioned.

There are three more developments I wish to note before leaving this opening nineteenth century narrative: photography, spectroscopy, and the beginnings of acoustic technologies. The simplistic science discovery narrative arises in part due to long established conventions. The "Great Man (few women)," Great Idea, "First Published" history itself has a long history, but that is rarely the way discoveries occur. This is certainly the case with photography, perhaps the first major, triumphal technoscience imaging invention. We have already noted that Ritter's use of silver chloride, a salt which turns black in the presence of light, is a negative imaging agent. Yet, even before Ritter, Johann Heinrich Schulze, a German physician, had begun to produce *images* in bottles of the solution which he exposed cutouts of words and shapes as early as 1727. He did not do any *fixing* of the images, since as soon as constructed, if one shook the bottle, the image would disappear.

The *camera obscura* had a long history of imaging, going back to at least 3000 BP (Before Present) and Mo-Ti, a Chinese mathematician, and by the European Renaissance it had become a major toy and tool for reproducing images of high verisimilitude.[7] Such "proto-cameras" remained in popular use in the nineteenth century and numerous individuals had the idea of a fixed image using such a device, for example, Tom Wedgwood as early as 1800 had the idea for a practical photography.[8] It was Joseph Niepce who first experimented with silver salts placed on a varnish film which produced stable images, and then with a *camera obscura,* invented what we would now recognize as he first photgraphic camera, 1822. Niepce continued to experiment, but with ill health and no truly successful simple film, he ended up selling his ideas to Louis-Jacques-Mande Daguerrre who was, finally, able to make plates of copper with silver salt coatings for the first practical photography. He published his process as a book in 1839 with immediate success and photographic machines began to be constructed throughout the industrialized world. Photography became an almost instantly ubiquitous technology, soon used for portraits, putting local painters out of business; entertainment, as magic lantern shows; military, with the American Civil War documented; and scientific uses, photographic apparatus connected to telescopes, common by 1840—a *triumphal technology.* As already noted, Roentgen later used photography to fix his x-ray results at the end of the century. Indeed, by the end of the nineteenth century, photography had become the essential recording technology for scientific claims.

There remain two more steps before leaving the nineteenth century: I return to the *camera obscura,* the optical toy and tool which in a sense launched early modernity both in art and science. I have elsewhere shown that since early modernity, seven or eight variations on the *camera obscura* were developed as types of imaging technologies for most of science's visual hermeneutics.[9] The earliest version also known as a "pinhole camera" yielded what I call *isomorphic images* in the sense that objects imaged retained their shape, color, and configuration. As noted, these images projected on a screen or wall inside the camera, were inverted and reduced from three to two dimensions, but they retained isomorphically the other visual-spatial features. Relatively early experimenters inserted circular lenses into the pinhole and by the eighteenth century were able to focus and manipulate the aperture to attain more resolution and brightness. But, as this chapter has indicated, Newton's imaging of the color spectrum and even Herschel's detection of infrared occurred with a pinhole and *with the addition of a prism,* but the opening remained circular. However, Thomas Young, as early as 1803–1804, began to vary the aperture opening by making it into a double slit opening. And when he projected light through this new shaped aperture, the result was a *wave interference pattern.* He used this evidence to argue for the wave nature of light (at this time some argued that light was particulate, including Newton with corpuscular theory, others argued that it was wave-like finally in late modernity, under quantum theory, it became both). Later, following Young's version of the double slit aperture, *interferometry* which shows wave interferences, developed. Note, now, that a wave interference pattern is nothing like either the white light being projected, nor like any object—it is rather a pattern which I call *nonisomorphic,* or unlike what is being imaged.[10] This will become clearer with the next aperture variant, the *single slit* and/or *diffraction grating* aperture.

At this point one can note that I am introducing two features to my narrative which have the advantage of making things clear, but which exceed a strictly historical and chronological perspective. All of us have the advantage of contemporary hindsight—we are denizens of the twenty-first century with its postmodernism. And by using the radical variations of postphenomenological theory, I can also show how each variation upon an instrument or technology produces a difference which is detectable and experienceable. Both of these heuristic moves help form insights into the technological embodiment of science in technoscience, but they also stimulate a certain unease to which I shall later turn.

For the moment, return to the earliest variant upon the aperture of the camera: a pinhole or circular lens variant. Both Newton and Herschel added a prism—itself clearly nonneutral and transformative of the resultant image. The prism separated white light into constituent spectral colors—for practical purposes here, "rainbow colors," ranging from red to blue to violet. Newton himself noted that his spectrum shape was ovaloid and that there was a sort of blending of colors, that is, no clear and distinct color separation. The same would have been the case with Herschel's spectrum. Each was, of course, interested in a different phenomenon: Newton thought that the phenomenon was white light, which his device divided into colors—so he used the spectrum as evidence for white light composed of multiple colors. Herschel, using the same setup focused upon areas beyond the imaged spectrum, to discover infrared light beyond light.

If, now, we return to the master narrative of science history, instrumental traditions also continued. Many scientists continued to experiment with optics and light, and so here I turn to a rather long set of experimental variants which eventually led to what today we recognize as *spectroscopy* (another variant on the *camera obscura).* This history spans the nineteenth century and both the multiplicity of names and associated Nobel Prizes is complex—many scientists were experimenting with this variant of the *camera obscura.* As early as 1802, W. H. Wollaston had been able to resolve the color spectrum into finer resolution which showed more distinct color lines, but by 1814 Joseph Frauenhofer first varied the aperture of the camera into a narrow, slit shape and later into a multiplicity of slits called a *diffraction grating*, which unlike Newton's ovaloid and ambiguated color spectrum, showed a clear set of color lines, neatly separated and resolved ("Frauenhofer lines"). Herschel and H. F. Talbot also heard about these experiments and replicated them. Finally, by 1859, Gustav Kirchoff and Robert Bunsen produced a practical spectroscope which produced the distinct color lines used in analytical spectroscopy. However, the most revolutionary insight, which resulted from the slit/ diffraction grating aperture variant, was the recognition that color patterns were actually *chemical signatures* of gases. Leon Foucault was the first to discover in 1848, that the sodium salt spectrum of gases in his laboratory setup was replicated exactly in the color pattern found in the sun's spectrum after which the chemical composition of stars could be made. So now from Newton's interest in the composition of white light, spectroscopy led to the chemical signatures of stars.

Readers by now will have noted that this narrative which covers so many technoscience discoveries of the nineteenth century actually remains dominantly within the visualist and optical centered history of science praxis and focus. So now, only at the end of this narrative do I turn to an acoustic focus. Yet, even more so than with the visualist technics of the nineteenth century, the acoustic technics are more anticipatory than not. The acoustic technologies of the nineteenth century show the same multi-discoverer proliferation as spectroscopy with one marked difference. The early—roughly first half of the century—remains largely within what could be called *resonance* or induction acoustic devices, not related to the theme of the electromagnetic spectrum which pervades the visualist technics already noted. However by mid to late century, soundwaves (rather than "ray" emissions, the preferred visualist terms) began to be discussed and developed. Recall in passing the medical acoustic technologies which improved diagnostic procedures—the progression of various hearing tubes to the stethoscope, 1816, which by mid-century with its tubes and more complex "horn" device had come into use. Much experimentation and variation belonged to this simple resonance acoustics—for example, wealthy people with multi-story dwellings, had internal communication systems (1849) which worked like a complex, long distance stethoscope. Bells connected by wires between floors alerted servants that a call was pending, then voice messages spoken through a tube system issues the command or request for a given service.[11] Interestingly, Jeremy Bentham at the end of the eighteenth century (1791) in his panopticon which was an early total surveillance prison design, usually gets

comment concerning how all prisoners were open to visual surveillance—but in his design he included a tube acoustic surveillance system
as well. Other longer distance nineteenth century sound systems included sound induction via railway tracks, and prior to wireless communication, over wire systems such as two-wire telegraph systems. Unlike
the tube systems in homes, the longer distance sonic communications
could not convey voices, but used Morse codes transmitted over wires.

Retrospectively, I would like to note that even into the nineteenth
century *there was no technology which could isomorphically image sound compared to isomorphic optics (telescope and microscope)* which isomorphically
imaged planet satellites to plant cells for early modern science. As Galileo
claimed "anyone" could recognize the mountains of the moon through
the telescope. Except for relatively short distances, all acoustic imaging
had some degree of *nonisomorphism,* although ambiguously. For example,
wired telegraphy sends its signals in terms of on/off clicks as in Morse
code. While this sounds *like* the sounds of the transmitter device and thus
could be called minimally sonically isomorphic, it is the sound of off/on
current disruption and the meaning of the message remains unknown to
anyone who has not been trained in Morse code, thus hearable ("read-
able") only to the trained listener. Thus, more like spectroscopy; only the
skilled reader of spectral codes knows which chemicals are being imaged.
Voices, musics, and other auditory phenomena were not yet sendable via
wire. This drastically changed in the last half of the nineteenth century.

When it comes to the telephone, the history of attempts and inventions is complex and controversial—literally dozens of persons tried
voice transmission technologies. The "lover's telegraph" a toy using a
taut string between two cans which served as diaphragms was long
known and voices could be heard over a short distance. But a telephone
which transformed soundwaves into electric waves did not develop until
the mid to late nineteenth century. Who should take precedence remains
filled with controversy. The standard history, or at least the history dominant in Anglophone countries, credits Alexander Graham Bell and Thomas Watson, whose 1876 patent was granted with this acoustic technological invention. But a contemporary, Antonio Meucci, a non-English speaking Italian immigrant made claims for an 1871 version and is still claimed
as the inventor in Italophone countries. And in what was to be a long
court battle, Elisha Gray also had produced a similar device and claimed
primacy. Eventually, the courts decided in favor of Bell. (The term, "telephone" was actually coined earlier in 1860 by Carl Gauss.) But with the
telephone the first isomorphic acoustic imaging of recognizable voices—
soon music and other sounds—occurs. Watson clearly recognized Bell's
voice in the message, "Mr. Watson, come here . . ." on March 10, 1876.
Technologically, just as hearing aids are more complex than eyeglasses,
so the telephone was to the telescope. I shall not here belabor the technical details, but the telephone needed a diaphragm which could change

sound vibrations into waves—in Bell's earliest versions these were liquid (mercury) sound produced waves which would produce motions, picked up by a wire or stylus which through an electrical device produced a current which could be sent by wire to the receiver, again with a diaphragm placed in some variant of a hearing horn. Note that physical motions, waves, could be translated from sounds and then again back into sounds. In short, sound as waves get physically produced by a diaphragm, turned into an electric current, reproduced as waves from a receiver diaphragm to a listener. Bell had several advantages over his competitors—he was trained in acoustics on the technical side, and had an education which included elocution training, a family with much deafness, and he, himself, studied speech training. Bell simply knew more about the world of sound than any of his competitors. His machines were tweaked to produce high analogue quality. The telephone, like photography earlier and x-ray machines later, caught on rapidly and the world rushed to be "connected."

The second invention which produced isomorphic acoustic imagery was the closely following phonograph, in 1877–1878. The technological world shrinks in the sense that its inventor is Thomas Edison, inventor of so many important-to-modernity items—like the incandescent lightbulb, only now going obsolete. Edison knew of Bell's work and Edison had been working on both the telegraph and telephone and wanted a machine which could repeat messages, thus a recording device was needed. Well aware of how waves could be used, he fixed on an even simpler shape transformation—a revolving cylinder, covered with tinfoil, onto which needle-impressed indentations could be made. His cylinders revolved and the needle made the impressions on tinfoil. His first machine was a two needle contraption, one needle for recording, the other to replay. Early cylinders did do the job—his first message, "Mary had a little lamb . . ." came back clearly, they reported. This event is usually dated August 1877, but he did not apply for a patent until December 24, 1877. It was granted in February 1878. Tinfoil cylinders were only good for a few plays, so experimentation for better recording surfaces began early. Waxes were tried, but soon bakeable materials began to be used. Early cylinders had only two minutes playing time for messages or musics, such as molded cylinders with closer tracks, by 1901 they could record four minutes of message or music. Here, I note, an interesting moment in art-technology history. Musicians soon "tuned themselves" to the instrument—four minute songs or performances became common and continued well into the programming on radio in the following century!

Initially, Edison fantasized many uses for his recording machines, running from Dictaphone to message and music recording, to various toys, memory files for families, and the like. The dominant uses, of course, soon became voice and music reproduction and, later, to become an infrastructure of technologies for the radio. With both the telephone and the phonograph, we now have the auditory equivalent of early modern optics—but neither is close to the breakthroughs implied by the electromagnetic spectrum.

As the end of the nineteenth century loomed, the explicit role of the electromagnetic spectrum emerged. The history of science recognizes James Clerk Maxwell as the theorist of the spectrum. He theorized the existence of the electromagnetic spectrum in the 1860s, recognizing that electric and magnetic waves existed across a vast spectrum—and he was the first to recognize that visible light must also belong to this continuum of waves. His contemporary, Heinrich Hertz performed experiments in 1886–1888 to evidence these waves and measure them. And he also measured the frequencies of sound waves. Neither were aware of the limits of the spectrum and although at the very end of the century some began to experiment with *radio waves,* which are at the long end of the spectrum, *gamma waves,* at the short end were not yet known (Ernst Rutherford discovered gamma waves in 1903).

The technology which utilized long waves (roughly a kilometer long) was the *wireless telegraph,* the immediate predecessor of the *radio.* Telegraphy, first in the form of a two-wire loop system, later in 1837 when metal plates were placed in the ground, forming a one wire system were common in the early century. Thomas Edison, Aleksandr Stepanovich Popov, and William Preece, actually did some successful wireless transmissions in 1880, 1892, and 1895, but Gugliemo Marconi got the first patent in 1896, and had a working machine by 1899. Marconi based his work on that of Maxwell and Hertz and by the end of the century in 1898 began to transmit wireless signals longer distances across the English Channel and in 1901 Cornwall to St. John's, Newfoundland. Much early wireless communication was shore to ship. The "sounds" transmitted were simple— telegraphic clicks, not yet voice or music.

Radio waves turned out to have some surprises: in certain respects, these waves were very much like x-rays. They were invisible, unfelt, and unperceived,traveled at the speed of light as other EMS waves, but due to atmospheric effects could travel great distances in curvilinear paths around the earth (Cornwall to St. Johns signals). And radio waves could also penetrate various kinds of solids. But until radio and even more powerful *radar,* the penetration of interiors was not yet a recognized nineteenth century EMS phenomenon.

Clearly, what can now be called late modernity was the beginning of global interconnectedness. Radio, oceanic cable, spanned the world. Retrospectively, these then new media appear to us to be of very poor quality with low noise to signal meaning ratios, and similarly with early cinema, lacking visual quality—yet in their own time they were enthusiastically received and both hearing and seeing was, in effect, "believing."

NOTES

1. Don Ihde, "Science has always been Technoscience," forthcoming in Jay Foster ed., *Continental Philosophy of Science* (Bloomsbury Press).

2. Isaac Newton, "Letters," in E. P . Bowles, *Galileo's Commandment*(New York: W. H. Freeman, 1997), p. 184 ff.

3. Peter Galison, *How Experiments End* (Chicago: University of Chicago Press, 1987). Galison is one of the preeminent philosopher-historians of science to make a turn to practice. He traces the development of "neutral currents" in twentieth century physics and recognizes that science practicioners develop instrumental traditions.

4. Betty Holzmann Kevles, *Naked to the Bone: Medical Imaging in the Twentieth Century* (Reading: Addison-Wesley, 1997), p. 19.

5. *Ibid.,* p. 20.

6. *Ibid.,* p. 21.

7. The camera obscura is a very ancient optical device. I am fond of my classical *Encyclopedia Brittanica* entry from 1929, Vol. 4, "Camera Obscura," p. 658 ff.

8. Peter Pollack, *The Picture History of Photography* (New York: Harry N. Abrams, 1977), p. 15.

9. Don Ihde, *Experimental Phenomenology: Multistability 2nd Edition* (Albany: SUNY University Press, 2012) pp. 155-170.

10. *Ibid.,* p. 162.

11. On my fiftieth high school reunion, our class toured the Seely Mansion in Abilene, Kansas, the home and factory of a patent medicine millionare. Built in 1904, a bit later than the tube acoustics nineteenth century arrival, this house was fully equiped with tube acoustics and had some original Edison lightbulbs. It was a trip back in time! Dwight Eisenhower, who grew up in Abilene, once delivered ice to the icebox of the owners.

TWO

Imaging, A Second Scientific Revolution

The narrative of chapter 1 traced, within the limits of the nineteenth century, the beginnings of breaking the limits of human perceivability as related to the *electromagnetic spectrum,* a major science theory breakthrough of that century. Yet, my narrative was also narrow, focusing upon what is bodily perceivable in relation to the technoscience of the times. Before returning to that theme, I would like to hint at what else was going on—in short, the *completion of the first Industrial Revolution.* At sea the world was globalized by sail and then steam connected by ocean spanning communication cables, economically related by burgeoning supra-national companies. I include here an anecdote—in 2007, I received a request to review a manuscript for Wesleyan Press, *Barbed Wire: An Ecology of Modernity,* by Reviel Netz, a philosopher-historian who produced a highly imaginative cultural history and philosophy of three closely related technologies—barbed wire, the telegraph, and railways. Note how closely they resemble each other. Each employs strands of iron or steel; each is supported by a series of wooden posts or ties; and each accompanies the other in the marches across all industrialized continents uniting and separating various things and domains. That image lingers as a powerful symbol of nineteenth century land industrialization. More, each technology is transformative toward a new future, very different from its premodern past.

Here a return to Marconi's wireless telegram is appropriate. In one sense it superceded the wired telegraph lines stretching across the plains. But it also needs one more step to become the acoustic equivalent of the telescope; it must become a *radio.* A radio is isomorphic; it transmits the voice and music. As noted in chapter 1, voice transmission begins in the twentieth century—1900. A Brazilian priest, Roberto Landell de Moura transmitted a voice broadcast in 1900; the early recognized the value of this technical possibility and traveled to the United States to apply for a patent (granted in 1904). In passing, it can be noticed that the radio was, more than most inventions of the time, an international phenomenon. Marconi was Italian, Jagadish Bose was Bengal Indian—an early transmitter of electric telegraphy, with experiments with what became known as a crystal radio—and Landell de Moura, Brazilian. Moreover, radio was contributed to by scores of early twentieth century inventors.[1] It was also both a wartime and an entertainment acoustic technology from the beginning. Early, mostly ship navigation and communication was entailed, but with aircraft, navigation use was immediate and both radio and wireless telegraphy were ubiquitous World War I technologies. By the 1920s, radio stations dotted the industrial world. BBC began broadcasting in 1922. With radio now able to communicate voice and music, there were three distinct technical means to convey sounds isomorphically. Although, for contemporary listeners this may seem strange: if one listens to an old gramophone or radio broadcast the ratio of "noise" to "signal" is very poor—yet just as early viewers jumped in their seats at early *cinema verite* so did listeners proclaim how clear Caruso sounded. My generation grew up on radio; we listened to "The Lone Ranger," "The Green Hornet," and in my case I became a devoted listener to the Texaco Metropolitan Opera to the mystification of my Kansas farmer father. And this phenomenon shows how a radio technology reversed a recording technology's constraint as well. Whole opera broadcasts could occur rather than the maximum four minute recordings of Edison cylinders.

Radio, however, was but one early acoustic first step in what I term the "second scientific revolution" which more fully occurs only from the twentieth century—the revolution in scientific imaging. Marconi had already discovered that antennae could be directed both to pick up and to direct signals. Early operators of radio also became familiar with *interference.* Listening to radio signals, particularly during stormy weather, produces static sounds on the radio. These were directional and varied with strength dependent upon distance. They were radio communicated analogues to listening to a thunderstorm in the mountains. Radio was the early isomorphic acoustic imager; radar, its spin-off, became the echolocation acoustic technology.

Although many persons noticed such static phenomena, Karl Jansky, an engineer with Bell Labs began to recognize that some signals were also coming from the radio spectrum of the heavens. Here there lies an irony; the very blockage of radio signals by the ionosphere which allows radio to be a long distance communication, from the outside prevents most radio signals from penetrating the atmosphere from space. Jansky did discover a regularity to what did come through, a source located in the Milky Way, now called Sagittarius A, a phenomenon which he confirmed and published as a discovery in 1933. Jansky and others located other sources, those which today we associate with previously unknown astronomical phenomena such as quasars, pulsars, and masers. To optical astronomy, this comes as a surprise because many of the sources are not optical, nor did they emit light. Eventually, the cosmic background radiation was also discovered and a Nobel Prize awarded to Robert Wood Wilson in 1978. In short, radio astronomy, itself discovered accidentally, was the first breakthrough into a totally nonoptical astronomy. Phenomenologically, such an acoustic technics was analogous to hearing: directionality and surroundability were dimensions of auditory space. But since astronomical sources did not correspond to optical sources, radio astronomy revealed sources beyond light, beyond optics.

If we now return to a more earthly technology, clearly within an acoustic analog, it is *radar,* the technological counterpart of *echo-location,* and its underwater counterpart, *sonar.* Both these technologies send sonic *ping* signals which when they hit a target, bounce back to the receiver at an—at first—listening station. Those familiar with these technologies are aware that in usual ocularcentric fashion, science replaced the auditory signal with a visual display screen or is otherwise visualized. For example, voice patterns are oscilligraph in form, and in many of today's digital synthesizers which produce musical sounds, a visual display is added to show wave shapes. What remained of listening stations and skills for radar and sonar through much of World War II, today are vast multi-screen displays at Air Traffic Control Centers or in submarines, ships, and the like. The acoustic technologies, however, function like animal echo-location and can determine shapes, directions, and speeds of objects in the air or undersea. However, recall that the technologies utilize radiation from the electromagnetic spectrum. By the end of the nineteenth century most scientists thought the spectrum to be a continuum which could possibly be infinite. However, with the discovery of gamma waves, the electromagnetic spectrum (EMS) began to be thought of as continuous, but bounded by both upper and lower limits. That remains today's notion, with radio waves at kilometer scale and gamma waves at nanoscale at the Planck limit.

As with the narrative in chapter 1, I shall take account of the vast proliferation of technical inventions in both acoustic and visual imaging which provide the base for what I am calling a "second scientific revolution" such that the world looks and sounds very differently than it did in early, and even late modernity. As noted previously, from the discovery of light beyond light, sound beyond sound, and then using various forms of radiation—alternatively called "rays" or "waves"—from the EMS, the nineteenth century had already produced both isomorphic and nonisomorphic image technologies. Photography dominated the first; spectroscopy, interferometry, and x-rays the latter. And the small step from wireless telegraphy and acoustic technologies of telephone and recording devices produced isomorphic auditory perceptions.

The twentieth century was as frenetic as the nineteenth in terms of expanding imaging technologies—here mixing visual and acoustic, I will do a very rapid survey:

- Acoustic technologies play a stronger role in the twentieth century. Radio plays a significant role and by the 1920s is ubiquitous, even to the present.
- Radar and sonar, with beginnings as early as 1904 with Christian Huismeyer and Robert Watson-Watt in 1915, plays a major role in World War II in the 1930s and 1940s. There is a contemporary irony here, very recently Japan has decided to begin to sell military technologies, now some seventy years later and one of its technologies is a highly refined anti-missile, anti-aircraft radar, aided by a laser target device to make location and distance even more precise.

- A spinoff—at first used for materials analysis by Reginald Fessenden, 1915—led to ultrasound applications such as medical ultrasound.
- The USSR causes a major impact in 1957 with the launch of the first orbital satellite, *Sputnik*. The United States, first responding with a failure a few months later, launches Explorer I in 1958 to open the Cold War Space Race. By the end of the twentieth century there are 2,200 earth-orbiting satellites. The range of imaging includes photography—including infrared and ultraviolet, which shows features such as plant and crop growth, thermal imaging, and radar, indeed imaging all along the EMS.
- Close on the heels of the satellite expansion of imaging, is the opening of solar and beyond solar system exploration. There have been almost as many space probes as satellites. I mention here only a few of the best-known instruments: The Hubble Space Telescope is probably the best-known and has yielded the most dramatic images: the Chandra x-ray source shows, for example, the pulsar with its axial radiation jets of the Crab Nebula; Cassini has produced art-grade images of Saturn's rings and moons; a whole series of vehicles on Mars (Mars Explorer, Sojourner Truth, etc.) continue surface and subsurface explorations searching for water and possible life; Venus probes have penetrated the toxic cloud cover to map Venus's surface—the list could go on and on.

The above schematic list of twentieth century imaging technologies is far short of a comprehensive list, but I would here like to comment upon several aspects of these technologies which account for their revolutionary qualities. First, as noted from chapter 1, the discovery of the EMS provides the framework for these technologies. A radiation continuum from long radio waves to nano short gamma waves is such that different instruments can be constructed to image the type of radiation by what I shall call "slices" of the continuum. Photography, for instance, can be used to image within the optical part of the spectrum—and, with only minor filter modification, this can be made to extend into infrared and ultraviolet ranges. Thus, as used in satellite photography, the light range can exceed the human visual perceptual range. Astronomical and satellite uses are perhaps the easiest to cite because the instrumental development is of high sophistication—and widely publicized, from classroom materials to coffee table books to television documentaries.

Yet, with each slice, there are also horizontal limits. One can photograph an x-ray image, but photography does not produce x-ray images. Thus, in satellite imaging, the Chandra X-Ray source can detect and image x-ray emissions from celestial sources and there can be an x-ray mapping of such sources. The same applies at different energies and frequencies with gamma waves or other distinctly different parts of the EMS. In the now three decades in which I have researched imaging technologies, it is interesting to note that while many disciplines use the same instruments and parts of the EMS, they do so with very different styles. For example, in both astronomy and medical imaging, multiple variants are used to image individual scientific phenomena. Here I will contrast a brain tumor in medical radiology with the Crab Nebula in astronomy. Radiological variants in brain imaging include MRIs and fMRIs, which are magnetic resonance imaging machines. These magnetically "vibrate" brain cells and can be calibrated at different frequencies for different results. But if one wants a more complete range of images, one also uses a PET scan which uses atomic level processes; and one can add a CT X-ray scan which uses computer tomography combined with x-rays for yet a third image variation. Then, using digital tomographic processes, one can produce a very clear 3-dimensional image of the object tumor. This is just what a surgeon most wants, that is, a detailed 3-D image of the pathological item to be removed. I call this a very postmodern technique. It relies upon multiple perspectives, but is topographically combinable into a 3-D composite image. It is also a multi-instrumental process which, in practice, is akin to applying "instrumental phenomenological variations."

If we switch to astronomy, the same multi-variant imaging can be performed—one can image the Crab Nebula with a range of technologies, including optical, stretching to infrared and ultraviolet, include gamma waves, radar waves, etc. The result will be to show the globular 3-D outline of the nebula. But if instead of this composite image, one wants to discern the inner structure of the Crab, then using only a "slice"—in this case the x-ray frequency—a spinning pulsar with radiation jets shooting out rays from the poles appears, and only appears, at these frequencies. Many astronomer friends described to me how they considered the best astronomy imaging technologies to be just these "slice" machines.

In both these examples, however, there is a process different from any prior to the twentieth century and which is produced from *computer pro-*

cesses. It is the process whereby all data can be transformed into image, and all images made into data.[2] This revolution in imaging capacity can be simply illustrated from the proliferation of imaging used in space or distance imaging. I have previously referred to Cassini, the Saturn imaging satellite which delivers startling close-up images of Saturn's rings. Historically, Saturn's "protrubences" were noted even by Galileo. But his telescope was not yet powerful enough to resolve these shapes into rings which occurred only when Huygens developed a more powerful telescope in 1654.

Cassini carries cameras onboard for close-ups. Remotely controlled by radio signals, these cameras take brilliant images—but to be returned to an earth station by radio, the photographic gestalts must be "reduced" to data-bits, transmitted, and then after arrival, reinverted into gestalt images of the rings and moons. This same process is implied in all *simulations,* the composite tumor, and the slice x-ray images referred to above. As will be shown below, not only scientists, but performance artists have adapted this new imaging capacity for far-reaching artistic productions as well. The earliest uses of the data/image processes were in early computer simulations such as the Manhattan Project of World War II. Early imaging was largely restricted to charts, projections, and simple imaging; today, digital processes of high quality range from movie to modelling productions.

In the previous chapter both spectroscopy and interferometry were noted. These two optical processes depended upon variants of the aperture (slit, double slit, diffraction grating) combined with prisms to produce what I call nonisomorphic images. Each class of instrument continued to be developed into the twentieth century—but with an interesting difference. Prior to the twentieth century light sources for these variants upon the *camera obscura* were natural light, such as the sun, or flame variants such as candles or various gas lamps. First each of these imaging technologies, with electric lighting, but then with much further refined light sources, could produce radically new effects. I will here note only selective examples:

- Spectroscopy became practical in 1859, later with beam sources such as electron and ion beams developed into *mass spectroscopy* a primary instrument for analytical chemistry. Its use in astronomy earlier had already produced chemical signatures of star types, but from 1899 and the discovery of "canal rays" produced by ion beams (note the nineteenth century language which still obtained) highly individualized chemical compounds could be identified. Two of my favorite examples are the identification of a particular volcanic obsidian from Western Pacific sources which turned up as trade objects in Indonesia or the far Eastern Pacific with dated sites indicating a trade path millennia old, and the specific stomach contents of Otzi, the 1991 Italian Alps Iceman, mass spectroscopically identified as red deer and mountain goat plus Eikorn wheat bread as part of his last meal while alive.

- Laser light, called *coherent* light, which is monochromatic, narrow beamed and is beam active for very long distances, was first produced by Theodore Maiman in 1960 in the Hughes Labs. Today, there are some 55,000 patents for laser light, including devices to read CD discs in play and to detect voice produced vibrations from surveillance targets (acoustic uses). In yet another *camra obscura variant,* laser light can be used to produce through a diffraction grating, an image of photon spin patterns.

- Fast following lasers, holograms, or 3-dimensional imaging, simultaneously invented in 1962 in the USSR by Yuri Denisyuk and at the University of Michigan by Emmett Leith and Juris Upanieka. Such 3-D holographic imaging is yet another variant upon the *camera obscura* technologies.

- Acoustic imaging at ultra- and infra-sound devices are also largely twentieth century in origins. Ultrasound (frequencies higher than 20,000 Hz) and infrasound (frequencies below 20 Hz) are the auditory equivalents of infrared and ultraviolet, sounds above and below human perceivability. Already noted is the relationship of infra- and ultrasound detection to radar, sonar, and of medical ultrasound. Although some techniques were used as early as World War I to detect locations of artillery through infrasound, useful detectors were produced by 1960 by Vladimer Gavreau in France. I shall return again to the animal perceivability of infra- and ultrasounds below.

Here a climax point has been reached. From chapter 1 and the nineteenth century, the EMS has become the framework for the vast continuum of radiation emissions, ranging from the more than kilometer waves of radio to the Planck constant nanowaves of gamma emissions. And, although I have done this through samples, the different parts of the EMS are both instrumentally detected and projected through the varieties of contemporary imaging technologies. Note, contrarily, that there is no single technology which does this to the full EMS, but through computer tomography both composite and data/image inversions can take place.

I will not undertake a separate chapter for the twenty-first century — although new and radical technologies are emerging with the same frenetic pace as noted in the first two chapters and I shall note the most interesting of these in subsequent chapters. Postphenomenologically, this now places us in the peculiar contemporary position of a full recognition that there is a very vast range of radiation emissions which we do not, and cannot, *directly* experience. First noting light beyond light and sound beyond sound, but later noting that a far vaster range of EMS wave

phenomena simply are not (directly) experienced at all in bodily perception. *But, also postphenomenologically, I would claim that we do experience-
perceive the entire range through technological or instrumental mediation.* This
includes Roentgen with his x-rays, the radio astronomers with their dark
signals, and such experience reaches into the new sounds of synthesized
music, the songs of "Caruso mice," and many other yet unidentified
sound beyond sound singers. Thus, if science in praxis is technologically
embodied, so also is science itself implicated in *human embodiment.*

NOTES

1. Bose's radio was a crystal radio, using a crystal for station location. We had one
of these in my boyhood home, which used a probe to find stations while scratching
across the crystal, and poor quality transmission could be heard through headsets. I
today regret that this radio was sold at my parents' estate auction at the end of their
lives.

2. Peter Galison, perhaps more deeply than any other philosopher of science, has
recognized both an image and a logic tradition within physics. This, together with the
recognition of contemporary imaging to invert and reinvert data and image is described in detail in Bruno Latour and Peter Weibel, *Iconoclash: Beyond the Image Wars*
(Cambridge: MIT Press, 2002), Peter Galison, "Images scatter into Data, Data gathers
into Images," pp. 352–359.

THREE
Animals and Robots

Although the emphasis to this point has clearly been upon technologies, particularly those developed around the discovery of the EMS in the nineteenth century, and on the other side upon human embodiment and bodily perception, there have been asides to nonhuman, animal perceptions. In this chapter I want to make a quite deliberate turn, mostly new to postphenomenology, which elevates two different types of variations into consideration—animal embodiment and perceivability deliberately varied with technological or even robot sensory analogues.

Animal studies have burst onto the scene with revolutionary force only in the last few decades. This is the case with the demise of mostly reductionistic and behavioristic paradigms for understanding animals. Today—and only for a few decades—it is acceptable to speak of animal "minds," "cultures," and even of "technologies." Readers will understand that in my case animal "technologies" were early interests. But, unless they were more familiar with my autobiography, they might not have understood that growing up on a farm in Kansas also meant that animals have always been a close part of my own experience. Indeed, I rode a horse to the one-room country school; domestic animals were always an intimate part of farm life; and pets have always been at least quasi-family members although until my mother reached more serious levels of rheumatoid arthritis debilitation, my father would not allow any in the house itself. Later, he relented and my mother had both a dog and parakeets as house companions. Farm human-animal relations were clearly not scientific, but based upon folk beliefs and practices. For my father there were "bad" animals—for example English sparrows, hated grain eaters, were particularly low on the scale, as were the traditional predators such as coyotes, and rodents such as mice and rats. "Good" animals included mostly feral cats who inhabited our barn and granaries, along with one of his favorites, bullsnakes, which also eat rodents. Dogs and horses were particularly valued and felt to be intelligent, and I learned much on my own about horse and dog communications, which in later life extended to a wider variety of wild animals ranging from bullfrogs to pond turtles to chickadees.

Academically, more narrowly and scientifically, the recent revolution in animal studies may be heavily credited to the rise of female and feminist thinkers.[1] Donna Haraway's *Primate Visions* (1989), was one pioneer study. As Haraway points out, mid-twentieth century female observer-researchers began to find significant roles in primatology. And once in place, the kinds of questions asked, tied to a different style of observation, transformed first primatology but today whole stretches of animal studies in biology. *Primate Visions* was one of the early feminist critical works and a later survey of reviews showed that most professional primatologists reacted negatively, but hisorians of science and science studies scholars were positive. I think today, one-quarter of a century later, the verdict is in—basic notions about everything from reproductive strategies to animal cultures, technologies, and personalities have drastically changed. I will look briefly at three examples: Haraway, Jane Goodall, and Sarah Blaffer Hrdy.

Haraway's *Primate Visions* was something of a manifesto. She took on the masculinist traditions of dominant male, Victorian family models, behaviorist determinism, and on into the Cold War mentality of experimental animals and provided a deconstruction. This she followed up by taking account of the new female primatologists who began to raise new questions, make new observations, which led to significant changes in how evolutionary reproduction was framed.

Jane Goodall (she and I are the same age) entered the field in 1955, under the tutelage of Louis Leakey, the eminent archaeologist-paleontologist. He helped provide her with scholarly, and eventually PhD (Cambridge 1962), credentials. He was convinced that females made for more careful observers with better notes, but was also aware that females were also accepted with less aggression on the part of male primates. Males sometimes regarded human males as competitors and would attack. Leakey basically pushed Goodall into her lifelong study of Chimpanzees in the Gombe Stream National Park. Goodall subsequently revolutionized our understanding of chimps—here are a few selective results:

- Goodall's style of observation, early criticized, was personal. She named her animals and recognized individuals by features and behaviors.
- She was the first primatologist to recognize chimp technologies—termite probes which were fashioned and preplanned for enough of them for a full meal; clubs to kill small antelopes, and was the first to publish articles on such weapon-based interspecies hunting;

pounder and anvil nut crackers. More recently, primatologists have observed female chimps make spear-pointed sticks to spear bush babies in tree cavities, thus making both sexes hunters. Today, more than 290 tool practices, distributed among some 57 distinct chimp troops are recognized. Interestingly, while some tool traditions can be shown to go back thousands of years, there is no evidence that there is technology transfer between troops.

- She recognized patterned grooming and social behaviors which helped stabilize social relations and hierarchies.
- She, with others, recognized distinct communication calls and behaviors, such that distinct warning calls for distinct predators (snake, eagle, tiger) were interesting prelinguistic behaviors.

One can easily see here how this approach to animal studies diminished the distance between humans and animals and how minds, cultures, and technologies could now be seen to belong to animals as well. One indication about how the field itself changed is the book *Baboon Metaphysics: The Evolution of a Social Mind* (2007) by Dorothy Cheney and Robert Sayforth. This is a detailed study of a baboon troop, which shows how intricate all of the behaviors of recognition must be for the troop to function.

I shall now turn briefly to Sarah Blaffer Hrdy, who is dealt with by Haraway, but who has a reputation and many publications in the field on her own. Hrdy, a Wellesley philosophy major, went on to her PhD on the *Langres of Abu* (India) in 1975. Hrdy came to her observations with a different question: the dominant interest of male primatologists seemed to be in successful conception, in short, reproduction was accomplished at conception. Hrdy changed perspectives to one of how do female langurs best preserve progeny once born? There is no successful reproduction if progeny does not survive. And from this follows a paradigm shift—the strategy is to get males not to kill offspring, but to feed them. Although this oversimplifies matters, what Hrdy found amongst female Langurs was a strategy of promiscuity and persuasion to help the feeding of offspring, and eventually this framework appeared to better describe reproductive behavior strategy.

Interestingly enough, today this shift from the dominant "Victorian" male of Haraway's displays, to the recognition that even amongst presumed monogamous birds (swans, eagles, etc.), it turns out that DNA tracing shows that biological paternity does not always equate with caretaker paternity. (Aware of this, in the forty-plus years I lived in my harbor front house in Poquott, my observations of swan fights between invading bachelors and resident caretaker mates revealed precisely this battle.)

If Haraway and the many feminist scientists have shifted the paradigm for understanding how reproduction occurs in much of the animal world, then the next step for introducing animal variants into postphenomenology and its understanding of embodiment must return to the terrain traced since the discovery of the EMS in the nineteenth century and its subsequent technologization since. As noted, today there are both detection and projection instruments which span the full continuum of the EMS—and one use for such technologies has been to detect and record the newly discovered perceptual capacities of animals (including insects). Today, based at the least upon documentaries, popular science publications, and television, almost everyone will be aware that our animal relatives often have light beyond light and sound beyond sound capacities. That is, many animals can visually perceive into infrared and ultraviolet ranges and auditorily into infrasound and ultrasound ranges. I will begin with selective examples which have come into scientific knowledge in the time frame noted.

One of the first animal groups to be discovered with infrasound capacity were the cetaceans or whales. The US Navy accidentally began to hear strange underwater sounds in the late 1950s—eventually these sounds were identified as sounds of whales, including the most musical, the humpbacks, but extending to virtually all whales and dolphins. If 20 Hz is the lower limit for human focal hearing, many whales may drop to 10 Hz. The 1960s whale songs had become popular on commercial recordings and were often integrated into other musics, and even became part of a popular campaign to "save the whales." Historically, ancient mariners likely also heard—but did not identify—whale songs as strange sounds from the deep, but identifications were often mythical as in siren songs, mermaids, or other fantasy creatures. Much whale singing reaches into twenty-plus Hz and thus can be heard without special equipment.

Of course whale songs, although perhaps the most seductive of ocean creature sounds, are only part of what today is recognized as a noisy ocean: shrimp *clicks*, fish *grunts*, and other animal sounds are detectable, but mostly overwhelmed today by shipping noise and submarine communication traffic. Today, after several decades now of study, one can recognize that ocean mammals have at least three kinds of acoustic systems: clicking sounds relate to echo-location, both for recognizing submarine shapes, but also for identifying prey; songs, or melodic sounds which today are recognized undergo annual patterns of change. Sea mammal music is periodic and highly sophisticated and obviously relates to both communication and sexual display. Other less melodic, but long drawn out soundings are probably more related to other communication patterns and probably also navigation. Infrasound whale sounds can travel thousands of miles. I will add, that other notably intelligent sea life, octopuses, cuttlefish, and squid also are sensitive to infrasound.

There is, in recent times, an unfortunate side effect largely from the human caused increase of ambient noise from commerce and machines in both urban and ocean environments. Studies show that urban birds have had to both raise the frequency levels and volume of their calls in urban environments, compared to rural environments. And recent dolphin studies show that they, too, have had to increase the volume of their calls to be heard by their peers—with the need for more oxygen to power these calls and thus there is a health stress associated with a noisier ocean.

Remaining with infrasound, many land animals also perceive these sounds which, I would claim, are "Mereau-Pontean" in conveying meanings such as the recognition of danger and various communication meanings. Many large animals have this perceivability: elephants, hippos, giraffes, rhinos, okapi, large cats including tigers and leopards. But, interestingly, also some relatively small animals also perceive and use infrasound: cassowaries, rock doves, pigeons. Of the latter, flying birds, it is thought that infrasound perception may be used in navigation (along with magnetic sensitivity which we do not seem to have). Although my list is suggestive, animal investigations are still underway and surprises occur with considerable frequency in the science magazines of the day.

If now ultrasound perception and emission is considered, some of the same animals reappear on the list, including whales and dolphins, but now other echo-locators such as bats. Recently it has been discovered that mice and rats communicate and some sing at ultrasound levels, such as various frogs, insects including crickets, katydids, and surprisingly, sloths. Ultrasound expression includes musics—mostly related to sexual courting behaviors—communications and warnings, navigation and echo-location. Equally surprising is the lack of primates on this list, although gibbons can vocalize at soprano level sounds. All the above are auditory perceptions below and above human horizons.

One of the most recent additions to singing mammals is the male mouse. Females also communicate at ultrasound, but so far do not exhibit song behavior. What came as my greatest surprise, however, is the standing stance (Figure 3.1) taken by these "Caruso mice!"

One final animal acoustic example, this time is at the level of very small insects. This week's *NY Times Science Times* reports on a parasitic ant nest beetle which has developed the uncanny ability to mimic the most important identification sounds of their victim ants. Both the ants and beetles make sounds with their legs rubbing against ridges on their bodies. But, the beetles have learned to make three different sounds which mimic ant identification calls for three distinct ant castes. (The investigators used highly sensitive microphones over long periods of time for acoustic observations.) Thus, even at the micro-level acoustic communication can be complex and evolutionarily rich.[2]

If shifting for the moment, visualizations are considered—first in infrared and ultraviolet ranges—then many insects and birds appear. Both these borderline color visualizations play important roles in pollination and search-and-find navigational capacities of the animals involved. The lowly Monarch butterfly turns out to be highly sophisticated—al-

"Caruso" Mouse   Ultrasound Singing

Figure 3.1.   Caruso Mouse Singing Ultrasound

though it has ultraviolet sight ranges which help navigation during clear days, when clouded it also has magnetic capacity, and for its 4000 km migrations can follow magnetic lines.[3] However, before I leave what now is a significantly large number of animal perceptual capacities which exceed those of humans, I want to turn to two examples of perceptual sensitivities, which so far simply do not occur with humans, or with very poor degrees: thermal imaging as perception and magnetic sensitivity, which plays a special role in embodiment orientation and navigations above. Both of these unusual forms of perception are also robotically developed and thus show the utility of the animal/robot variations I am recommending.

One reason for considering animal variations lies with evolution which is usually very slow change-over-time. In this process, variant solutions to similar or same problems often occur. For example, bat echo-location occurs through click-sounds emitted by the bat. The discrete sounds hit targets and return for the bat to hear and to perceptually determine size, shape, and distance through binary hearing and time differences. But different species of bats emit their click-sounds different-ly, most emit them through their mouths and vocalizations, but some emit through their noses. (Humans do a parallel variation with flute play-ing; most flutes are played via mouth breathing techniques, but there are also nose flutes with nose breathing techniques.) Bat echo-location, and aquatically by dolphins, is clearly "imagistic" in the sense that they can both acoustically differentiate extremely small differences in targets. Dol-phins can acoustically tell the difference between tennis ball size balls if one is only one-quarter inch in diameter different in size. Similarly, those animals that can visualize into infrared or ultraviolet dominantly do this through visual perception and eye physiology. But in the case of animal thermal perception I am about to take up, a different evolutionary variant has been arrived at—I refer to the thermal perception of reptiles, most keenly that of pit vipers or thermally perceiving snakes.

Returning to Herschel and his experience of infrared—he "felt" the heat of infrared in that part of the spectrum from which he could see no light, thus one could say he had something of a thermal perception. But it yielded not color, let alone shape or other visualizable features. Take your own thermal perception—I am sure you can feel when you get close to an outdoor fire or an indoor wood burning stove, or even a countertop burner. But humans cannot thermally do fine sensory-like shape discrim-inations. The rattlesnake, a pit viper, can apparently perceive a nearby mouse as a kind of heat image, detect its motions and the like. But inter-estingly, this is not done by eye. Pit viper evolution has instead opted upon a different sensory form—there is a pit, near its eyes, with heat sensitive cells which in genuine perceptual fashion produce a figure/ground thermal gestalt of the mouse. In contrast, a cat—perhaps feral—which has very sensitive night vision, would pick up the same mouse shape visually. Thus, the evolutionary variant for the snake is, in effect, a different kind of sense organ, whereas the cat simply has a more ex-tended visual organ. But in both cases shapes are apparently "sensed." This recalls my difference between focal perception and "felt" field per-ception in listening to rock or feeling infrared heat.

If we now turn to a technological set of variants, there is also a close parallelism: everyone is familiar with "night vision" devices. The evening news depicts soldiers in Iraq or Afghanistan wearing monocular goggles on top of their helmets. These yield a green glowing image of what is in front of the soldier. Green glow, I note, is quite common for EMS related imaging. However, there are two variants of such devices—one is usually called "night vision" and relies upon enhanced light, the other more usually called "thermal vision" pulls in infrared color, but both yield shaped images of the scene. Some of these devices are also used in, for example, helicopter surveillance, such as that which showed the moving body image of the Boston bomber inside the boat in which he was hiding and projected on the evening news. Thermal imaging was also used—unsuccessfully—in the search for the two upstate New York murderers from Clinton Prison. Some reports indicated that the wooded area in which they were hiding had too many heat images from deer, bear, and cattle and thus did not get a human enough image. Technological thermal vision also shows heat shadows, for example, if vehicles or airplanes have just left positions, their heat shadows remain. It should be easy to see how closely animal and robotic or technological variants map upon each other and how both mediate what would not be possible for naked human perception to be mediatedly perceived.

The next step I wish to take relates to an animal, and a technological, set of variants even more remote to ordinary human perception. In this case I refer to *magnetic sensitivity.* In the animal cases, a considerable number of animals—migrators usually—have magnetic sensitivities. That is, they apparently perceive the earth's magnetic lines. Physiologically these animals are usually found to have forms of iron or iron oxide located within their bodies. In the case of birds, often these are located in the head, near the beaks. Homing pigeons, but all sorts of migratory birds, apparently have this capacity. A quick switch to technological variants, of course, include compass or compass-like detectors. In the animal case the experiential dimension is bodily perceivability; in the compass cases, one "reads" the compass display. One cannot be sure what the animal magnetic sense is, or how it is experientially shaped, but it is clearly focally sensed. For humans, "reading" a compass is more *hermeneutic* than an isomorphic bodily in relation. One anecdote: I used to try to teach novices how to steer a straight course with my sailboat. I quickly learned that to try to teach straight course steering by reading a compass was a poor tactic. Invariably, the double attention—today's multitasking—ended up with an "S" course. It was much easier to have the novice steersperson fix his or her sight upon some distant object and thus not overreact to the compass needle.

While it is clear that current animal studies are far from comprehensive and are still at an early stage—for example no one knows what the full range of animals with magnetic sensitivities may be—studies increasingly add to the magnetic phenomenon. Insects such as bees and termites evidence magnetic sensitivity. In these cases navigational utility is obvious — bees to find nectar sources; termites to correctly connect tunnels in habitat building. Birds, some turtles, bats, and in some recent cases accidentally discovered, cattle and deer appear to have this sensitivity as well. In 2008, observers scanning Google Earth noted that in Europe many grazing cattle lined up in north-south orientations — unless the graze area was close to electric power lines in which case orientation was random. Hynek Burda, who had been studying magnetic sensitivity in blind mole rats, came across the cattle and deer grazing observations and did a scientific sampling (8,510 cattle photographed at 308 locations) and, unless near powerlines, the north-south orientation obtained.[4] I had previously read about this phenomenon and since I frequently take regional train trips between cities and universities in Europe, began to keep track of grazing cattle and, to my surprise, the north-south orientation did dominantly obtain for my informal sightings!

We now have one set of samples from animal studies, which show animal perceptual sensitivity into regions beyond those of humans (infrared, ultraviolet, infra-, and ultrasound) and have noted at least two forms of perceptual capacities either barely present to humans (thermal) and magnetic capacity possibly totally lacking in humans. Postphenomenology, as readers know, is amenable to pragmatism, including that of John Dewey who made his interrelational ontology modeled after the evolutionary organism-environment interactions of both human and animal life. In short, what the animal variations show is a different pattern of bodily-perceptual organism-environment interaction. I now reach the final point of this chapter. If we return to technologies, there is in particular among roboticists, an interest in what today is called *biomimetics*, or the use of "animal design" for technologies.

The idea of biomimetics is quite simple: animals are taken as displaying models, solutions, and results, honed by millions of years of evolution, which can be applied to human problems — problems related to engineering style design. The term was coined and used by Otto Schmidt during his own doctoral work in the 1950s and applied to a trigger mechanism in squid in 1957. But in practice, the observation of animal characteristics and applied to engineering problems is much older. Leonardo da Vinci (1400–1500s) did particular observations and drawings of bat and bird wings to understand flight capacities. Similarly, the Wright brothers did observations of pigeons and their flight and wings in designing their own first successful flight technologies.

There are and have been hundreds of such applications, increasing in contemporary life. For example, Francis Bottenberg[5] has studied the extreme ability of geckos to move upside down with their micro-hairs providing holding power. Adaptations for dry medical tapes, and other types of adhesives, have followed[6] Velcro, which uses hook and loop mini-technologies, patterned after plant burrs. Light refraction from but-terfly and other insect wings are adapted to solar panel improvements for higher efficiencies, and the list goes on and on. Today, MIT has a biomi-metic lab—much of its output relates to robot motility, such as the series of Cheetahs which have been highly publicized. Ironically, Hubert Drey-fus, the iconic critic of old style artificial intelligence has had strong indi-rect impact on this field by claiming that motility is in fact a harder problem to solve than calculations. Roboticists have taken this challenge seriously and insect or multiped modeled robots are showing more suc-cess in recent years. Similarly, small flying robots, some the size of large insects or small birds, have been derived from biomimetic researches.

I have concluded this chapter with a turn to motility, again because one deep theme of postphenomenological research relates to bodily skills, embodiment. Embodiment is also part of human motility whether in ordinary action or in specialized and skilled actions such as dance, gymnastics, and other specialized human actions. New skills, such as those entailed in laproscopic medicine or the remote control of robotic devices are emergent embodiment skills today.

NOTES

1. Don Ihde, "Feminism and the Philosophy of Science," *Postphenomenology: Essays in the Postmodern Context* (Evanston: Northwestern University Press, 1993) pp. 117-136.

2. Carl Zimmer, "Mimicking a Way In," *NY Times,* Tuesday, July 21, 2015, p. D.3.

3. "Butterflies steer with magnetic compass," *Science,* July 4, 2014, 345 Issue, p. 43.

4. Daniel Cressey, "The Mystery of the Magnetic Cows," *Nature,* 10, 11 November 2011, p. 1038.

5. Francis Bottenberg, presented her findings at the Society for the Social Studies for Science but results were not published to date.

6. *Ibid.,* p. 1038.

FOUR

Is There a Bat Problem for Postphenomenology?

Thomas Nagel's famous article, "What is it like to be a bat?" was published in 1974.[1] In context, it was an argument against physicalist reductionism. Nagel was arguing that conscious experience—subjectivity—could not be reduced to brain states, but in the process also concluded that we could not know what it was like to be a bat. That year was crucial for me in a phenomenological sense—I had been working on my book, *Listening and Voice: A Phenomenology of Sound* (1976, first edition, later a second edition, 2007). In the process, and during writing from 1972–1974, I was teaching myself to echo-locate. This was to lead to a set of auditory experiments which I used to teach my phenomenology students to also learn a degree of echo-location in the halls of Harriman Hall at Stony Brook.

Those familiar with the history of twentieth century philosophies will be aware that both phenomenology (Husserl) and pragmatism (Dewey) were busy trying to deconstruct early modern epistemology (Descartes and Locke, seventeenth century). At the turn of the millennium, *Nature* magazine invited me to write a millennium essay, which I did with the title, "Epistemology Engines" (2000).[2] In that essay I showed how both Descartes and Locke separately but deliberately modeled the birth of early modern epistemology on the *camera obscura,* a favorite optical device of the Renaissance and early modernity. This was to be the invention of "subjectivity" and "objectivity" which modeled the eye to mind to a homunocular subject in the box viewing mental events. Knowledge was a correspondence of an event inside and an object outside the camera box and with it experience becomes "subjective" and private.

As I reread Nagel, I wondered how far he had escaped the seventeenth century. Here is what I found: unlike Descartes, Nagel was gener-

35

ous to conscious subjectivity well beyond humans, "Conscious experience is a widespread phenomenon. It occurs at many levels of animal life, though we cannot be sure of its presence in the simpler organisms. . . ."[3] And he eschews the crudeness of the homunocular subject-in-a-box, "I am not averting here to the alleged privacy of experience to its possessor."[4] Although I worry that there is a vestige of that position in his insistence that ". . . the facts of experience. . . . are accessible only from one point of view."[5] And, finally, although he is arguing against physicalist reductionism, he gives back what he takes away by averting to brain structures which are different in bats and humans, thus making it impossible for us to fully imagine being a bat—their brain structure is different.[6]

The other side of the argument refers to the subjective nature of experience which in the article displays considerable ambivalence. Nagel's argument relies on how close bat subjectivity might be to human subjectivity and since echo-location is obviously more refined with bats, we apparently are not able to approximate or imagine into such a subjectivity (privacy again?). Nagel does avert to heightened auditory abilities of skilled blind persons in a footnote[7] but all this remains short of bat subjectivity. So, while I conclude Nagel has become more sophisticated than an early modern, he has done so only to a limited degree. But I now want to leave Nagel and turn to phenomenology and postphenomenology, using the bat problem as an index for a difference between these two versions of phenomenology.

If Dewey and Husserl began the attack upon early modern epistemology early in the twentieth century, Maurice Merleau-Ponty dealt it a *coup-de-grace* by mid-century. "This phenomenal field is not an 'inner world,' the 'phenomenon' is not a 'state of consciousness' or a 'mental fact' and the experience of phenomena is not an introspection."[8] How far is this from the "subjectivity" of a single alien consciousness of a Nagel bat? "To be a consciousness, or rather *to be an experience*, is to have an inner communication with the world, the body, and others, to be with them rather than beside them."[9] Here we have a quite different set of descriptions from those of early modernity. I am taking Merleau-Ponty as a classic phenomenologist who takes a motile whole body approach to experienced embodiment. It should be noted that already here phenomenology no longer fits into the primacy of subjectivity and consciousness. Embodiment is actional, more than mere consciousness, and ontologically inter-relational, in-the-world. Later I will point to Merleau-Ponty's equally suggestive postphenomenological anticipations through instrumental material means. But for the moment I would like to follow a trajectory which links what I shall call *connoisseur perception* to classical phenomenology.

Connoisseur perception is skilled, trained perception. Here are a few simple variations:

- Wine tasting is such a skill. There has been more than a fifty year tradition of a wine tasting contests between teams from Oxford and Cambridge universities in the UK. Chosen teams of student tasters take up rigorous training for several months before the competitive event. They learn to detect the wine variety (most oenophiles can easily tell the difference between a cabernet, merlot, malbec, etc.), vintage, region, and multiple distinct flavors related to the tannin, acid, sugar contents of wine materiality. This is trained, connoisseur perception, an embodiment skill.
- More ordinary—bird watching. There is a vast variety of finch and sparrow species, all of which are likely to be perceived simply as small, brown birds by the amateur. But with trained perception, one can soon tell the different markings on the bird's heads, throats, wing patterns, beak patterns, and the like—plus the distinctive songs which differ greatly among song sparrows, white throated sparrows, chipping sparrows, and the like, recognizable to connoisseur perception, the embodiment skill for appropriate phenomenological observation.
- Or, one of my own favorites is mushroom hunting. My wife and I recognize a varied set of Vermont mushrooms for warm weather eating: oysters, chanterelles, boletus, puff, etc., etc. We are amused that some guests are afraid to sample, but others find fresh trout and chanterelles a special delight.

Classical phenomenology, I claim, reaches these levels of experiential analysis. All the examples have been of unaided or unmediated perceptual-bodily experience—embodiment. And here I can now return to the bat problem. Auditory connoisseur perception can easily recognize a range of human echo-location. Nagel recognized some skills attained particularly by blind auditory connoisseurs—I have seen some documentaries of blind bicycle riders. These riders produce a mouthed click sound (clicks or sounds of short temporal span are best for echo-location. Longer sounds as in songs are different, see below). These riders are able to detect poles and other objects down to three inch diameters. In *Listening and Voice* I have a chapter on perceiving shapes, surfaces, and interiors, all aspects of human echo-location. One can auditorily determine spatial-material features. A shoe box in which one can place glass marbles, then replace these with gummy bear marbles, each of which sound different regarding shape and density. Pounding on walls can reveal where a stud is. A granite countertop sounds different from a wooden one, etc. All of this is material-spatial as is the more finely tuned bat version.

I also try to keep up with much frontier work in acoustic and auditory discovery and two bat trivia are worth citing: in Central America small tungara frogs are prey for night feeding bats. These frogs sing mating songs hoping to attract females—but the frogs can hear bat echo-location clicks and so stop their songs. But it is too late, because the sound waves from the songs have produced ripples on the puddles they frequent and the bat can "read" the wave pattern to find the frog![10] Or, a more interesting bat-flower interaction is that of certain flowering plants which change shape at night to form a reflecting shape which can return the echoes of bat clicks so they may be discovered and pollinated. This is another exquisite example of interrelational adaptation.[11]

These examples of connoisseur perception are all excellent variants upon Merleau-Ponty's notion of perceptual-bodily motility as part of the mutually human world interrelationality. I now want to begin the extension to postphenomenology which in two further steps *incorporates* two different types of bodily perception by taking "technologies" or specifically in this case instrumental technologies into *intentionality* or phenomenological interrelationality. I shall call what results *instrumentally enhanced* and *instrumentally translational perception.* Again, Merleau Ponty may be seen to have anticipated this move. I have previously referred to his example of an extended sense of embodiment or bodily-environmental experience with the feathered hat of a lady and of the blind man's cane. In both these cases the experience is mediated through a material artifact, and gives the human experiencer a perception of extended embodiment. Mediating instrumentation, however, gives a more precise example of this now enhanced and extended perception. Such extensions may be, in fact, geographically very distant. Robot warfare often has the pilots of drones stationed in the United States—Utah or Arizona—and yet these pilots have kinesthetic awareness of flight.

Or, one may find many examples of both visual and acoustic imaging which pushes all previous magnificational or amplificational limits. Electronic and scanning tunneling microscopes for micro phenomena, multi-lensed and compound telescopes for macro phenomena are now deployed. These make it possible for enhanced bodily-perceptual experience never previously possible and it is to this human-instrument-world interrelationship that postphenomenology can and does attend.

What I am here calling instrumentally translational perception is yet another mode of pushing horizons. I have frequently pointed out that all astronomy—until definitively in the twentieth century—was optical. Ancient "eyeball" observations were revolutionized by instrumentally enhanced, telescopic optics in early modernity. Trajectories of improvements led to bigger and better telescopes, but these remained limited to optical light. Until the accidental discovery of radio astronomy in the early twentieth century, radio emissions including the background radiation of the universe were unknown. But since radio astronomy is at least analogous to hearing, this is not yet what I am calling translational. Rather, emissions which are "real" but which exceed any of our senses, are the worldly phenomena which can now be translated. Reverting to animal embodiment, certain birds, some mammals, fish, and reptiles apparently can perceive magnetic lines within the magnetic field (as discussed in chapter 3). This perceptual dimension I do not believe humans have but with the animals who do have it, navigation, orientation, etc., can be embodied.

Much of what is happening, following the paradigm shifts in animal studies, is that the traditional vast human/animal gap in Western culture is today being rapidly eroded. We are finding ourselves both closer to our animal relatives, and yet different from them. The genetic revolution today places us much closer to animals with over 30 percent of our DNA identical to *c elegans,* a favored laboratory worm; 85 percent with our close dog friends; and more than 97 percent with chimps. Yet, our perceptual systems in good evolutionary organism-environment patterning is also different from our relatives—for example, our house cats who have much better nocturnal vision.

In technologies, however, there are many familiar instruments which can provide translated perception for us: metal detectors for finding coins on a beach, Geiger counters for detecting degrees of radiation, even compasses for navigating are all ordinary examples of such instrumental translation perception. Indeed, astronomically, we can now image radiation from nano-scaled gamma rays to kilometer length radio waves, thus making frequencies far beyond our normal perceptual capacities mediatedly experiencable. All of this falls within the purview of a postphenomenology open to technoscience effects.

To this point I have emphasized perceptual embodiment, from ordinary up through connoisseur to instrumentally mediated bodily perception. However, with instrumental translational perception certain different factors come into play. Postphenomenology does indeed pay rigorous attention to embodiment—with Merleau Ponty postphenomenology recognizes that human motility, whole body actional embodiment is involved in all human and even intelligent behavior. But with instrumental translational perception, whereby dimensions and aspects of phenomena not directly experiencable, different factors emerge. Here I return to the early development of my work in the philosophy of technology. Clearly the signature and best-known of this work remains my "Phenomenology of Technics." Its first publication in book form was in *Technics and Praxis* (1979), and then expanded and refined in *Technology and the Lifeworld* (1990). In the later form, the phenomenology of technics was reprinted in at least seven anthologies and to date has been translated into three languages. This work analyses human-technology-world interrelationships, of which *embodiment* relations constitute an important type. From the beginning I used scientific instruments as one set of examples, such as some referred to above. But a second set of these interrelations I called *hermeneutic.* As the term suggests, hermeneutics entails a specialized kind of interpretation. Let us take reading as a practice—reading is clearly a perceptual, bodily activity, but what is being perceived are inscriptions, texts, codes, or also visual displays as in science. All of this is still an embodied praxis—but with a difference which here I relate to *isomorphism*. Galileo clearly saw the moon through his telescope (instrumentally enhanced perception) but the isomorphism remains recognizable in spite of and through the previously unseen features of craters, seas, and mountains. But with what I called hermeneutic relations in both books, degrees of isomorphism begin to change. Instruments still refer to a worldly phenomenon, but begin to vary away from perceptual isomorphism. A dial or mercury thermometer refers to how hot or cold something is, but one does not "feel" how hot or cold it may be outside or inside an environment. The same is the case with most instrumental translational perception. My sailboat's compass is referring to magnetic lines, but I "read" rather than feel them as lines. Granted, much design is often directed toward what would be greater perceptual isomorphism. The x-ray part of the spectrum which shows the rapidly rotating pulsar at the center of the Crab Nebula, complete with its radiation jets from the poles, retains a spatial isomorphism for the radiation, whereas a spectrograph visual display would show only a type of "bar code" of spectral color bands. (I object to the common use of these conventions as "false color" since they are assigned for the users purposes. Assigned color or referential color would be better.)

A more ordinary use translational instrument is an oscillograph-like display which transforms spoken speech into visual graph patterns. These instruments are used by various voice teachers to—for one use— teach how to reduce accents. Each sound shows a distinct visual pattern and if the accent speaker can see how the teacher's pattern differs, the accented speaker can learn to change the pattern by changing the sound. This instrumentation is also a good example of what I call science's visual hermeneutics. The habitual pattern in science is to transform all displays into visualizations—here sound patterns into visually graphic patterns. Even sonar, originally delivered with sounded pings, soon became a visual display. Instrumental translational perceptions, such as I have referenced here, may be either translations between sensory dimensions (sound to sight or other intra-sensory translations) or translations from nonperceivable dimensions into sensory ones (magnetic lines or x-rays to visualizations). These more radical translations are also ways in which new "worlds" become available to human experience. One indirect effect on postphenomenological culture has been that by analyzing and appreciating translational perception, postphenomenologists are shown not to be anti-science, but open to new directions of investigation.

To this point, however, the bat question itself has framed my discussion in a dominantly individual-experiential way. Thus, whether the above interrelational examples were without, or with, instrumental mediation, enhancement, translational forms, the reflexive role of experience remained that of a single, embodied experiencer. Yet, from classical phenomenology on, all experience has been understood to be *intersubjective.* That was one major point from Husserl's *Cartesian Meditations,* and is yet another meaning of Merleau Pontly's "being outside myself in the world." In my terminology, particularly as worked out in *Technology and the Lifeworld* (1990) intersubjectivity, sociality, is already included in *macroperception* and a *cultural hermeneutics.* I argued that such intersubjective experience is already experienced in embodied life. As a very simple example, take eating—we Westerners employ a skilled cutlery practice, knives, spoons, and forks. These technologies all have long histories: today's table knives go back to daggers; spoons, probably the oldest, go back to wooden and bone ones; forks, also from daggers, multiplied tines from one to two or more. And, we teach our very young—baby spoons— to begin to acquire the bodily skills of this use. And this use is embedded in conventional etiquette and coded significations. Setting the table means the fork on the left and knife and spoon on the right with the blade facing the plate. I well remember the slowness with which I learned a French restaurant convention on my first trip there. I could not figure out why waiters were slow to deliver "l'addition" until someone pointed out that I needed to lay my cutlery at an angle on the upper right side of the plate as a signal that I was done. Still using this example, I want to move to another insight of a postphenomenology. There is more than one way to have human-technology eating skills. Most of us who either live in pluralistic contexts or travel, become adept at the Asian counterpart of Western cutlery—chopsticks and spoons. I learned this skill during undergraduate days, once leaving my Kansas farm context with its meat and potatoes-plus diet, and learning in those long-ago days to eat chop suey and chow mein, I learned the use of chopsticks. Many decades later on lecture tours of China, I was pleased to hear many times a complement on how well I used chopsticks! I call this a *multistable practice* and from this humble example I now want to turn to the postphenomenological development of multistability.

My own first investigation into multistability was my first edition of *Experimental Phenomenology* (1977) which dealt with multistable visual perceptions. These were what then were usually called "ambiguous drawings," Necker cubes, duck-rabbits, face-vases, and the like (Figure 4.1). Much of the earlier work on perceptual reversibility arose from Gestalt psychology and its close relationship with early phenomenology. Almost all examples from these early studies were of two- or at most three- stability variations. Duck-rabbits and face-vases, old and young woman drawings were two stabilities, Necker cubes rarely, but sometimes were reportedly three (a flat two-dimensional variation was reported, usually interpreted as a result of observer fatigue). My use of rigorous variational analysis, however, showed the whole series of differently positioned Necker cubes and I developed a *five variant* version for each of the differently positioned cubes. Similarly, in several other cases I took what had been previously noted as two-variant forms to nine or ten variants (circles with squiggles). Finally, utilizing what turned out to be simple orientation variations, I transformed the famous duck-rabbit into a four-variant duck-rabbit-squid-Martian quadruple.[12]

I did not realize in the mid-1970s, that multistability was also beginning to be used in the sciences. I recently discovered that our university president, Jack Marurger, a physicist, had been working on multistable optics, with an entry in 1976, and then from the mid-1980s a boom has occurred in many sciences. In microchemistry, two atom layers of various materials have been found to produce multiple stacking patterns with 5<13 variants; galaxies have finite but many variants; and even infinities have 7.[13] What I understand today is that by a rigorous application of variational theory to various phenomena, unlike early phenomenology which discovered an invariant or "essence," I found that I was discovering multistablities. As it turned out, this phenomenon was especially useful concerning technologies and their cultural adaptations. Thus, later than *Experimental Phenomenology*, first edition (1977) I worked out the complex versions of Pacific compared to European navigational systems as multistable in *Technology and the Lifeworld* (1990), and a whole set of instrumental multistabilities in the second edition of *Experimental Phenomenology: Multistabilities* (2012).

Although with multistability we are rather far from the bat problem, there is one more reference back to the bat and Nagel I want to make. My suspicion is that Nagel's insistence that every subjective phenomenon be connected with a single point of view retains a vestigial privacy notion, it clearly is phenomenologically belied by the polymorphism of experience and the mulitstability of the simple multiculturalism of eating practices to the higher level more complex practices of navigation to the multistability of the material world which contemporary science is beginning to recognize. As Albert Borgmann has so nicely pointed out in *Postphenome-*

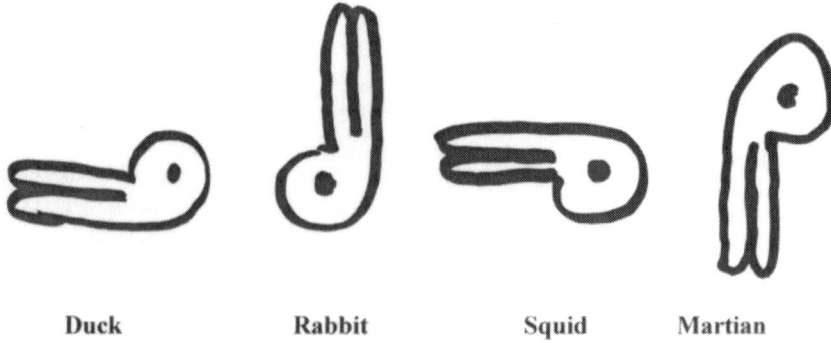

    Duck   **Rabbit**   **Squid**   **Martian**

Figure 4.1.   Illustration Multistable Doodle

nology: Critical Companion to Ihde, postphenomenology emphasizes multi-perspectivalism interrelated to multistability between humans and their world.

NOTES

1. Thomas Nagel, "What is it like to be a bat?" *The Philosophical Review* LXXXIII (4 October, 1974) 435-450. p. 1 [html pagination].

2. Don Ihde, "Epistemology Engines," *Nature* Vol. 406, 6 July, 2000, p. 21.

3. Nagel, *op. cite,* p. 1.

4. *Ibid.,* p. 1.

5. *Ibid.,* p. 2.

6. *Ibid.,* p. 2.

7. *Ibid.,* p. 8.

8. Maurice Merleau-Ponty, *The Phenomenology of Perception,* translated by Donald Landes (London and New York: Routledge Taylor and Francis Group, 2012. Original French publication, 1945), p. 59.

9. *Ibid.,* p. 99.

10. *Smithsonian Magazine,* 23 January 2014. p. 24.

11. *National Geographic Magazine,* "Call of the Bloom," p. 128–135

12. *Multistable Doodle,* Don Ihde.

13. Natalie Angier, "Multiple Infinities," *NY Times,* December 31, 2012.

2

Dimensional or "Case" Studies of Acoustic Capacities

FIVE

Technologies — Musics — Embodiment

Today, recorded music probably accounts for the single largest category of music listening. This chapter seeks to reframe the usual understanding of the role of that type of music. Here the history and phenomenology of instrumentally mediated musics examines prehistoric instruments and their relationship to skilled, embodied performance, to innovations in technologies which produce multistable trajectories which result in different musics. The ancient relationship between the technologies of archery and that of stringed instruments is both historically and phenomenologically examined. This narrative is then paralleled by a similar examination of the history and variations upon recorded and then electronically produced music. The interrelation of musics-technologies and embodiment underlies this interpretation of musical production.

Michel Foucault, in trying to convince us that premodernity had a form of knowledge, an episteme, which is now past and no longer makes sense, claims that the symmetry of resemblances which ruled the sixteenth century led to a conclusion that "there are the same number of fishes in the water as there are animals . . . the same number of beings in the water and the surface of the earth as there are in the sky, the inhabitants of the former corresponding to the latter . . ."[1]

We laugh or are amused — what sort of claim is this? Is it empirical? But if so, then, theoretically, we could go about confirming or disconfirming it by a count. Yet, what agency — the National Science Foundation, NASA — would support a grant to do such a count? We know from the outset that such agencies would not, although in today's episteme, one might be able to fund a census which would take a count of polar bears over several years to determine if they are entering endangered species levels, or do a census of whales to see if they have recovered their populations enough to be hunted. Foucault's point is that the thinking which relies upon such presumed symmetries simply no longer has any bite; its episteme is dead.

The purpose of this chapter is to undertake philosophical reflections upon recorded music, or as I prefer in a parallel to Foucault's epistemes, musics in the plural. I will begin my reflection with an attempt to locate us with respect to the many musics we may experience contemporarily and hint at something like the suggested census noted above. I shall first forefront the listener, the human who hears or listens to musics. But in parallel fashion, I shall also place the listener into a context where the technologies which mediate the musics are also brought under scrutiny. Clearly, there are indefinitely large numbers of such musics which could be listened to: performed musics—chamber, classical, rock, ancient music consorts, street performers, and the list expands and expands. Similarly, whatever types of performances might be listened to, such musics could also be recorded and thus listened to in the form of recorded music. Recorded musics are—we usually think—replicated musics. But here one must then also account for the plurality of recording technologies: iPods, Walkman, vinyl, CD, digital tape, Musak, radio. And again there is an indefinite list of technology which mediates and presents the recorded musics.

Yet, with an echo of the difficulty of determining if the number of birds equals the number of fishes, there is difficulty in determining how many listeners hear how many songs in how many ways. Yet there may be clues: gold records are those which sell 500,000 per run; platinum records are those which sell 1,000,000 per run. In 2006, in the United States, there were 30 gold and 16 platinum runs, thus equaling 31 million records.[2] Now, how many people listened to those 31 million records and how many times was each song heard? And don't forget to count all the downloads which also occurred in the same year from the multiple sources. While we do not know an actual number, we can easily surmise that the number must be very, very large indeed! But, to make it simpler, while yet remaining intuitive, let us imagine only listeners to recorded songs in greater New York on a given day, and then imagine the listeners to all the live music performances of whatever kind on that same day—from the philharmonic in Lincoln Center to the Peruvian flutists in Greenwich Village. I am willing to wager that the number of recorded presentations is simply much larger than the number heard in live performances by a very large magnitude. Here my point is that we philosophers, musicologists, performers, and other theorists may be coming to reflect upon recorded music very late. For if it is the case, as I suspect, that on a worldwide basis more listeners listen to recorded music than any other kind of presented music, it is sort of like a very large canary already escaped from its cage and now has grown too big ever to re-enter.

Recorded musics are, of course, technologically mediated musics. And, historically, recorded musics are quite recent arrivals upon the very ancient histories and even prehistories of other musical presentations. For the moment I shall not discuss scoring, which could be considered a sort of prerecording technique for preserving some kind of identity between the same piece played in different performances, and which also employs a type of "material technology" analogous to print, but as notations on a page. Nor shall I do more than sketch the very rapid history of recorded musics which go back only 130 years, but I do want to point out in bulleted blinks how fast and how diverse this technological trajectory has been:

- In 1877, Thomas Edison produces the first useable cylinder recording, first with tinfoil, later wax cylinders. These were mechanical devices recording sound waves physically, mechanically. Only a few plays are possible before the record deteriorates.
- By 1899 coin-in-box predecessors to juke boxes were already popular.
- By 1902, Caruso began to record, first with cylinders, later with discs, which began to appear in 1903.
- The year 1904 saw the invention of the diode, which made electrical rather than mechanical recording possible, but which did not become practical until 1919.
- By 1923, radio threatened to depress the reproduction industries, first with live, later with recorded presentations.
- The first stereo developments began in 1931 and magnetic tape followed in 1934.
- Vinyl, which reduced surface noise compared to the older discs, began in 1948 and the older '78s began to be replaced by '45s and '33s. Full stereo was available by 1956.
- Cassettes in 1963, digital CDs in 1978, and DAT or digital tapes came on in 1989.
- Then, with the 1990s came the proliferation of online and downloading copies in all the varieties now popular.

I want here simply to make two points: first, this 130 year proliferation must appear as a very rapid proliferation which also covers a wide variety of different technologies to record and reproduce musics. Second, it is a history, which while having ups and downs, clearly is one which now pervades an entire global economy, again evidence of our very large escaped canary.

I have now begun with technologies, which in the case of recording technologies mediate musics which in this first pass are heard by listeners. In short, I am relating here a material means of producing musics to experiencing humans who listen to these musical phenomena. Now, however, I want to shift to prehistory and begin now a long-range location for musics. How long ago humans began to make music remains unknown—but whenever music began, from the earliest human beginnings there were always already present human uses of technologies! This may seem like a strange claim, but only a few years ago there appeared in *Science* an article which refers to the chipped stone tools used by chimpanzees to crack nuts in a hammer/anvil fashion which go back at least 4,300 years BP (Before Present).[3] The sample found remains quite identical with contemporary chimp tool formation and use. The surmise of the article is that such simple tool use probably goes back at least to the common ancestor from which chimpanzees and humans split off! By current dating that is over six million years ago. And, for the physical anthropology literate, we are all familiar with tool uses by homo erectus, Neanderthals, and homo sapiens. Stone Age tools go back at least two plus million years. I would like to suggest that our common image here is one of a limited appreciation of the diversity of technologies by our ancient ancestors—we may think of Acheulean hand axes, or chipping tools, or maybe if we realize that "soft" technologies such as nets and baskets probably were also used, but all of these, we usually surmise, are for subsistence needs. That is, we tend to evoke a simple and basic existence for our predecessors. This may underestimate our ancestors.

What, then, should we make of a 45000 BP "bear bone flute" found in a site associated with Neanderthal humans in a cave in Slovenia? *Science* reported this find and the probable conclusion that this artifact was likely a flute as evidenced by the regular symmetrical shaping of the four holes which, under analysis, yield a tuning system for a diatonic scale.[4] And while such a musical instrument is not millions of years old, so as to compete with Acheulean hand axes, it probably does suggest that performed instrumental music occurred long ago in prehistory. In passing, note that we are now shifting the music scene from listeners—although they, too, were likely present—to also include performers. With performers, human embodiment actions come into play. The flute player must learn an embodiment skill which engages, in this case, the disciplined hand and breath motions which are mediated through the flute to produce music. We are now able to recognize a very basic relational ontology of instrument use. The human practitioner plays the flute to produce musical sound—I diagram this as:

Human — flute — music

This relational phenomenon is what phenomenologists call "intentionality." But in this case, it is an actional intentionality which is directed, mediated through a material instrument—a technology. A deeper analysis would go on to show that in the learning process, the shapes of experience change: first, struggles with playing the flute yield sounds, but they are not refined, gracile, "musical." But as skill is acquired, the flute is "mastered" in that it withdraws or becomes more and more transparent and the player is able to produce the sounds we hear as flute-music. This same process, which we can describe for the movement from novice to virtuoso performance, no doubt also characterized the experience and attainment of the Neanderthal flutist. The "woodwind," here "bone-wind," instrument permits the mediation of the human hand and breathing action into flute music, audible both by the player and any audience present.

We have now taken a simple look at recent recording technologies and then at a very ancient instrumental technology with musics mediated by different technologies over a very vast historical span. Now I want to risk reader dizziness by reciting what for me was a very formative occasion, a conference on musical improvisation at the University of California, San Diego, in 1981: I had been invited to be a keynote speaker at this conference which appeared to me to be of great interest—an interdisciplinary gathering of musicians, composers, humanities academics, and even one other philosopher, Daniel Charles from the University of Paris. I arrived realizing that I knew not one person from previous experience, although at social events I met folk who had read my *Listening and Voice: A Phenomenology of Sound*, which was, of course, the connection to this event and the source which motivated the invitation. But it was the improvisation workshops which turned out to produce the most interesting provocations.

I arrived at a workshop, a studio in which there was a grand piano, various traditional instruments, and a collection of distinctly nontraditional instruments. I tried to be unobtrusive and found a seat in a corner, self-consciously aware that my only performance abilities were abandoned long ago from high school to early undergraduate trombone playing days. But observer status was not an option—I was handed a "water horn" and commanded to participate. The water horn was a stainless steel container, partially filled with water, to which had been brazed a series of brass rods of different lengths around the perimeter of the vessel. I was handed a violin bow, and by now cacophony had already begun. Some players were hitting the open piano wires with hammers; others were turning anything in sight into a percussion instrument; virtually nothing was being played in a standard or traditional way and so, in the spirit of the event—which was simultaneously being recorded for posterity—I began to bow the rods on my water horn, and later to shake the instrument to get a gurgling sound. Extreme improvisation indeed, but this event took in my experience and memory as I realized that any instrument whatsoever had multistable possibilities just as I had earlier noted belonged to various other perceptual phenomena. This event, in 1981, occurred not long after my first book on the philosophy of technology (*Technics and Praxis*, 1979) in which I had initiated a long-term interest in the role of instruments as the material means for producing scientific knowledge. And, I claim, there is a parallel between scientific and musical instruments with respect to the already noted role of intentionality. The human action undertaken is mediated through the instrument to produce its result. If, in one case, it is the greater knowledge Galileo obtained through his telescope—of the craters of the moon, the satellites of Jupiter, the phases of Venus, in the other case it is the transformed sounds produced through the playing of the instrument, sounds which are as different from ordinary human vocal sounds as the sights through the telescope differ from those of naked eye observation. And to attain each production, in both cases, there is an acquisition of a skill which must be acquired to have a refined and high quality result. I shall return to this parallelism later for added analysis, but for now I want to return once again to an ancient prehistorical example of technologies which turn out to display a remarkable set of multistabilities which I find surprising.

My example is that of ancient archery: with few exceptions (Australian Aboriginals who developed boomerang technologies and some Southern Hemisphere groups who developed blow gun technologies) virtually every ancient culture developed archery, variations upon bows and arrows. But the style of use, the technical composition of the material artifacts, and the cultural contexts into which these technologies fit, display an amazing set of multistable patterns:

- The English longbow, constructed of yew or ash, long (2 m-plus) with long arrows, could be taught simply for use for the yeoman and fired with rapidity from a standing position. The firing technique called for the bow to be held at arm's length before the bowman, with the bowstring then pulled back and the arrow placed between the first and second fingers for release. Its effectiveness was demonstrated in the historical battle of Agincourt, where this type of archery overcame the crossbow archery of the enemy.
- From the East came the radically recurved, short (1.2–3 m) and composite bow used by Mongol horsemen who repeatedly invaded Eastern Europe. Clearly a bow of longbow length could not work on a galloping horse, and the composite of bone, wood, skin, and glue, radically recurved, allowed a smaller weapon similar power. But the firing technique was also different. Here the bowstring is held close to the face and bow pushed rapidly away with another form of quick fire, timed exactly to the gallop of the horse.
- A third "artillery" style of archery arose in China, in this case a long (2 m-plus), recurved bow called for the highest pull power in antiquity (matched only today with the compound pulley powered bows now popular). Here the firing technique included simultaneous push and pull, plus the use of a thumb ring to prevent injury to the thumb by the string. I was delighted to discover some of the

terra cotta warriors in the XiAn complex cast in exactly this posture on my first visit to China in 2004!

The examples above illustrate what I call multistability in the sense that the "same" technology takes quite different shapes in different contexts. In each case, the tensioned, strung bow can propel the arrow over distance with striking power, but the human skills take different shapes and patterns in using the also technically different artifacts. Does this seeming excursus have anything to do with music? My answer is "yes" and I shall try to open this line of inquiry by returning to the improvisation event with its playful exploration of performance variations upon traditional and nontraditional instruments.

Every archer could hear the bow string "twang" when fired. Could it then be "played"? We have already noted at least three styles of firing an arrow: bow extended and held still; string held still and bow pushed out; and double push and pull. Each of these variations, however, serve the same purpose, to fire an arrow. But in a new context if one holds the bow in a horizontal position instead, and "plucks" the bowstring—we are transforming the bow from its usual use, into a new use, as a sort of stringed instrument!

Anyone familiar with a history of instruments, or ethnomusicologists would know that there have been a wide variety of single-string instruments in many cultures. But I suspect not many know that there is also a tradition of actually using an archery bow as a stringed instrument. Apparently, such uses are common in Africa. According to even such a common source as the *Encyclopedia Britannica*, ". . . The San of the Kalahari often convert their hunting bows to musical use."[5] So, we are now off on a new technological trajectory or line of development. To produce a more interesting music, why not a moving fret? And, we notice, that the simple "twang" is not very loud or powerful, so why not add a resonator?

Again, just as with the hunting bow, there are a variety of types: "There are three types: bows with a separate resonator; bows with attached resonators; and mouth bows [all] evolved from the hunting bow."[6] In short, what I am doing with this set of cultural variations, is to show how an improviser creates a new stability or performance practice, using either a regular archery bow, or by adding very minor features and thus opening the way to developing stringed instruments. In my actual, original speculations—phenomenological fantasy variations—I did not yet know about what I cited above, but followed what I knew to be an exploration of different ways in which the human-technology actions could be possible as in the improvisation event. But as it turns out, this trajectory had already been followed. The above illustrations are of sub-Saharan African practices. But such practices apparently are not only contemporary practices. Another source, archaeologist J. D. Lewis-

Williams points out in the South African Archeological Society Newsletter:

> Bushmen recognize two kinds of music, vocal music and instrumental music. . . . the more personal instrumental music . . . is played . . . sometimes while walking in the veld, sometimes while relaxing in the camp. One of the most characteristic Bushman instruments is the musical bow, which is, in fact, an ordinary hunting bow. It can be played with the performer using his mouth as a resonator. When so used, the bow is more or less horizontal but can also be played with the stave vertical or semi-vertical. Then some performers like to use a calabash or other object as a resonator to produce more varied sounds as they tap the string with a stick. [7]

Then, to clinch the case, these modes of playing the bow (Figure 5.1) have recently also been found to be depicted in the traditional Bushman rock art found in the Natal Drakensberg area by Paul den Hoed and Justin Clarke.

This rock art dates back to 2500 BP. (I visited this area and saw some of this rock art in 1982 but did not see this particular depiction.) And, if this is not enough, note that a shaman depicted in the Trois Freres cave in Southern France with a dating of 15000 BP is shown, as now commonly interpreted as playing a bow in this same position.

With bows played in this way, both multistable variations appear, and in each different types of musics are produced. "Apart from adapted shooting bows, more specialized types of musical bows are widespread. Most are sounded by plucking or striking the string, but the Xhosa uubhu is bowed with a friction stick . . ."[8] Thus, the "same technology"—a bow—apparently fits two radically different trajectories, one of them musical. And this set of different trajectories is apparently also very ancient.

On the surface, it may seem that this detour into prehistory with glimpses of ancient "bearwind"/woodwind instruments, or of stringed weapons/stringed instruments may seem very far from the focus upon recorded music. But my point here is that, in human-techology interrelations, there may be found a set of multistable variations, and, from these, suggestions for trajectories or further and different developments. To fortify both points permit one more set of variations, this time those which may be found in a similar comparison between scientific and musical instruments.

It is probably not coincidental that the European Renaissance and early modern science both marked a period in which instrumentation began to proliferate in both art and science practice.[9] In music, this is a period which instruments are more and more used, compared to older a cappella and plainsong sacred music, to the increased use and experimentation with a variety of stringed, brass, woodwind, and percussion instruments. Indeed, our current orchestral instruments such as violins,

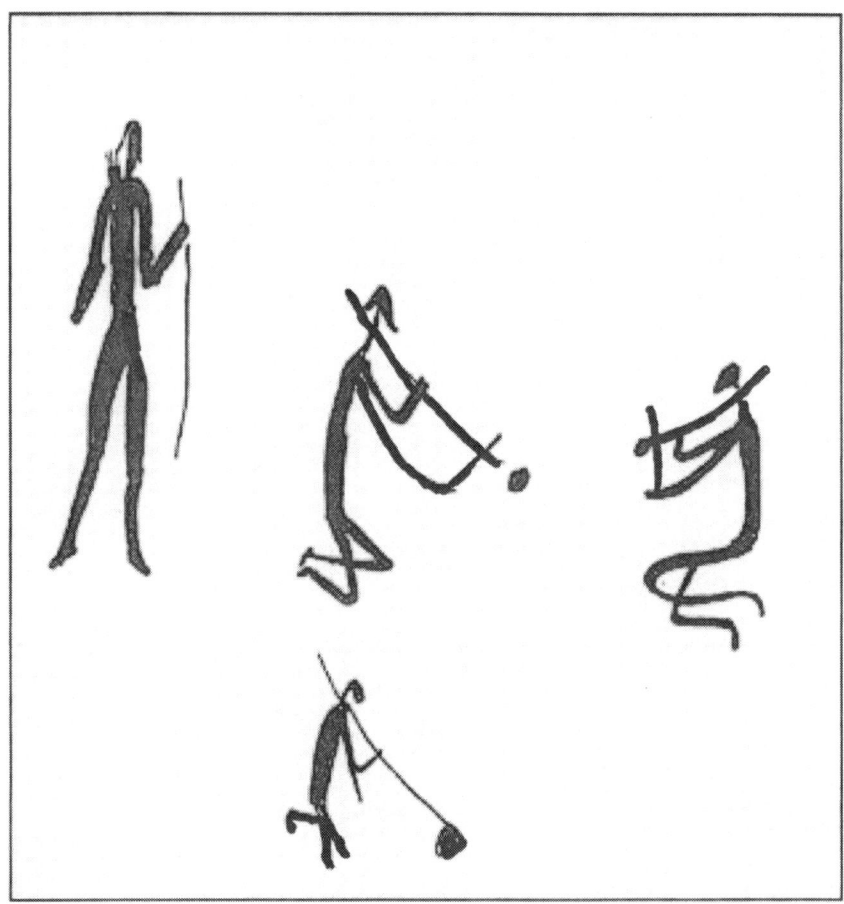

Figure 5.1.   Bushmen Playing Bows, 4000 BP

violas, horns, timpani, and the like, can all trace their development to this period. Musical instruments, however, produce sounds but many other Renaissance and early modern instruments were more related to optics and visualizations. Galileo, often taken as the paradigmatic figure for early modern science, developed both telescopes and microscopes utilizing compound lenses to magnify both the macroscopic and the microscopic phenomena of interest. I have referred above to several of his observations of previously unseen celestial phenomena—and these could only become visible by first recognizing the optical possibility of magnification, and then gradually developing the optical technology which both allowed greater resolution and greater magnification. Indeed, Galileo, hearing of Lippershey's 3X telescope, went on to produce nearly a hundred of his own telescopes, up to approximately 30X, which turns

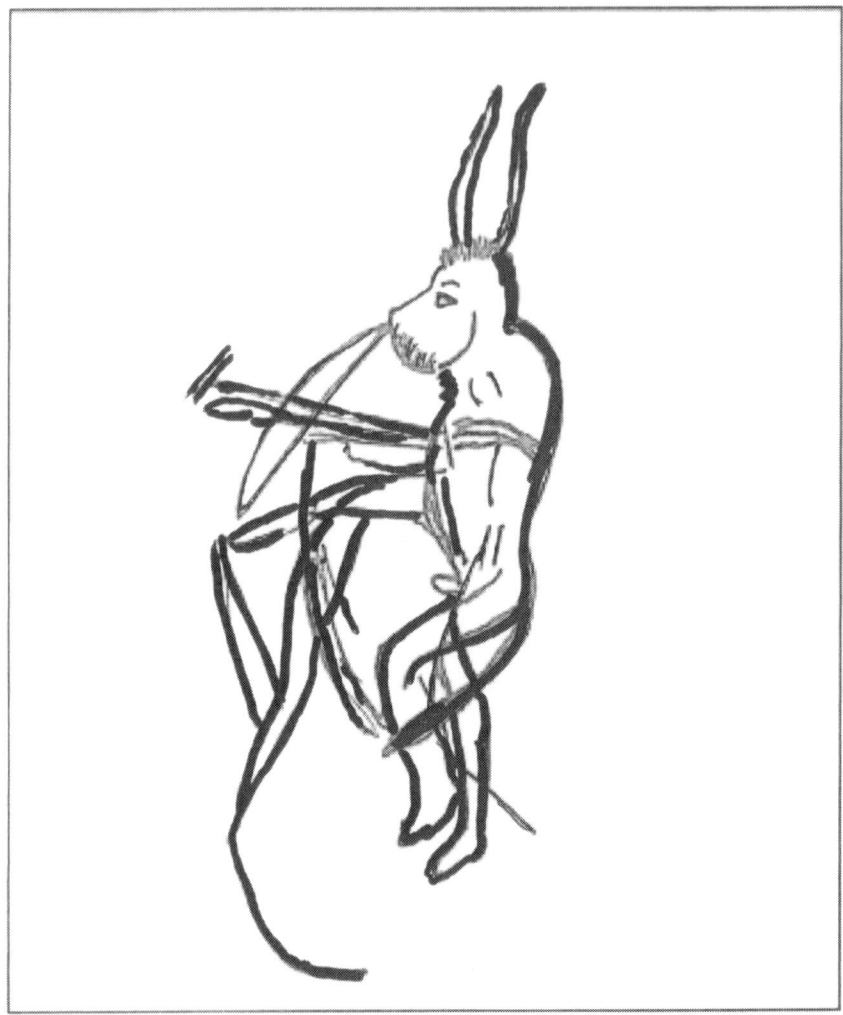

Figure 5.2. Shaman with Bow in Play Position, 15000 BP

out to be the limit for lenses without encountering chromatic distortion.[10] Thus, the limit Galileo reached was one which allowed him to recognize the "protuberances" of Saturn, but not resolve them into the rings with which we are now familiar. The trajectory, of course, is the line of development which recognizes in the material possibilities of optics, the possibility of greater and greater magnification and resolution, suggested in the very use of the instrument. The same following of a trajectory is, of course, also possible with musical instruments. Changes of material for stringed instruments, for example from gut to hair to wire or polymer strings, all allow different tonalities for the produced sounds. Nor should we forget that the human actions in "playing" or tuning both scientific and musical instruments also plays a crucial role in the output. Galileo insisted that in order to see what he saw, he needed to train the novice telescope user, not unlike the training which must go into producing a good tone and sound from any musical instrument.

There is, however, a crucial difference which may illustrate a quite different subcultural contrast between the history of scientific, compared to musical instrumentation. Once again, I revert to a somewhat imaginative variation for illustration: I ask—could you imagine a serious twenty-first century astronomer directing the graduate students in astronomy to try to make a new discovery of some astronomical phenomenon by picking up and reusing one of Galileo's telescopes? Yet, especially when I visit Europe, I have often been delighted to attend concerts given by groups who pick up and use ancient or early instruments to perform a chamber piece! And, even today, what violinist would turn down the opportunity to play a concerto using a Stradivarius or Guaneri violin? I am suggesting that there is lurking here a strong contrast between the instrumental traditions of much science and of much music culture. In general, I am claiming that there is an inbuilt progressivism in the adaptation to new instrumentation associated with science practice, but there can be an equally inbuilt romanticism which sometimes results in a preferred traditionalism associated with some music culture. This may also be evidenced by historical examples.

For example, Trevor Pinch and Karin Bijsterveld have long been interested in the way in which new technologies have been received within historical musical culture, and they have discovered what I shall call "Heideggerian moments." They note in "Breaches and Boundaries in the Reception of New Technology in Music," that technological innovations are frequently first decried, "For instance, the introduction of the piano forte was seen by some as an unwarranted intrusion of a mechanical device into a musical culture which revered the harpsichord."[11] And, again, when in the nineteenth century, ". . . key mechanisms and valves (such as found on today's woodwind instruments) were introduced to replace the traditional means of controlling pitch by the use of fingers over the individual holes. The new valves and keys were found to be easy to operate and facilitated the production of much more uniform and cleaner tones for individual notes [such a change met opposition from one, Heinrich Grenser, who complained that improving tone . . . by the use of keys was] . . . neither complex nor art . . . the real art of flute construction was to build flutes which would enable flutists to play whatever they wanted without the use of keys."[12] I call this a "Heideggerian moment" because the objection to mechanical valves and keys parallels Heidegger's famous rejection of typewriters in favor of the pen:

Human beings "act" through the hand; for the hand is, like the word, a distinguishing characteristic of humans. Only a being, such as the human, that "has" the word can and must "have hands" . . . the hand contains the essence of the human being because the word, as the essential region of the hand, is the essential ground of being human.

[After which Heidegger goes on to discredit the typewriter] It is not by chance that modern man writes "with" the typewriter and "dictates . . . into" the machine. This "history of the kinds of writing is at the same time one of the major reasons for the increasing destruction of the word. The word no longer passes through the hand as it writes and acts authentically but through the mechanized pressure of the hand. The typewriter snatches script from the essential realm of the hand—and this means the hand is removed from the essential realm of the word."[13]

One can easily substitute hand harp playing for keyboard piano playing and one can see the equivalence to the "Heidegger moment" in the resistance to changes in music technologies.

This resistance is perhaps understandable in the following sense. I have noted above that with any musical instrument which entails human bodily action and the acquisition of skills, presupposes long practice and development which the introduction of a new technology may disrupt. Moreover, any new technological development also enhances and simultaneously reduces some quality to the produced sounds which also thus changes the music. Yet, at the same time, new skills can be acquired and a new virtuosity may be attained. To those who complained about the "mechanical" output of the piano forte, who today would think of Ashkenaszy or Ash as producing "mechanical" sounds from the piano? It would be very hard for one who had spent years of hours of skill acquisition to simply abandon such skills for every new modification in instrumentation.

Only now am I ready to return to the focal topic of recorded music. Take note of the implicit trajectory in the short history of recorded music: the earliest cylinder records had very poor fidelity, could play only very short pieces, and could replay them only a few times. In spite of this, early listeners often gasped at how "realistic" the voices sounded. Yet, in the interplay of designers and listeners to this recorded music, it could be immediately clear that, if it was possible to improve fidelity, lengthen playing time, and increase repeatability, one should do so and the reexamination of the noted history shows that this aim was followed:

- Most of the early cylinders allowed only 2.5 to 3 minutes of playing time; this led to matching the musical piece to the recording time-apacity and "popular" tunes of segments of arias prevailed.
- By 1903, discs had begun to prevail, but still playing time was short. Thus, HMV Italiana's release of Verdi's "Emani" took forty discs.

- The leap from mechanical to electrical recording, 1877 to 1919, was irreversible, but background noise still posed a fidelity problem.
- Stereo, initiated in 1931, added depth to recordings and vinyl. In 1948, depth to recording and with vinyl discs combined to produce much higher quality recordings.
- The switch from analog to digital and the invention of the CD, paralleled by the amplifier tube to transister technologies produced changes in sound qualities which found loyalists who divided between listeners who preferred the richer tonalities of vinyl and tape recordings over the background noiseless, but crisper digital sounds.

This history, shorter than that of early modern optics, displays the same trajectory desires, here for greater sound fidelity parallel to the visual desire for greater magnification and resolution. In the process, however, something else also happens: the technologies for producing these results become much more complex and compounded and with this evolution there arises the possibility of greater manipulability. In order to return to recorded music technologies with a new perspective, I will here return to the music parallel and my final science instrument examples.

Early modern astronomy experimented with compound lens telescopes, all of the refracting sort (tubes with lenses and a focusing device), but, as mentioned, by the time one reaches 30X the fact that white light is made up of a spectrum of colors which refract at slightly different wave frequencies, meant that resolution would be poor and a chromatic distortion sets in. Newton, a century after Galileo, discovered this and reasoned that if one could use a parabolic mirror to refocus the different wave lengths, one could overcome the chromatic distortion—thus, the reflecting telescope which combines lenses and mirrors. Then also, as multistable variants were tried with optics—prisms instead of lenses producing spectra, twin slits producing wave imaging—more and more compound devices produced more and more previously unknown phenomena. A truly revolutionary step was taken, however, only with the discovery of digital processing utilizing computers in the twentieth century, which made possible many of the images now familiar within astronomy. Computer tomography makes possible the various manipulations which today show us everything from extra-solar planets to spinning pulsars.[14] Indeed, I have a close astronomer friend who claims that a telescope is no longer even considered an instrument; it is merely a light gathering device to which are attached the variety of "instruments" such as spectrometers, interferometers, etc.

I am now finally ready to return to recorded music and its technologies, but with a new framework for understanding, a framework which is no longer bound to taking recorded music as simply copies of, or representations of previous or simply performed music! Rather—and this is the crucial shift—the new combinations of technologies in a complex gestalt could themselves be considered to be a different instrument or instrument set. In my simplest examples above, such as adding a gourd resonator to a hunting bow, one changes and makes the instrument both more complex and yet more "resonant" for the musical sound produced. Admittedly, the much later addition of electronic amplification is a more dramatic change but one would not deny that the electric guitars played by rock musicians are simply different instruments—indeed one can hardly imagine a rock concert without extensive systems of amplification. And more subtle forms of amplification, often subtly "hidden" now, are part of the musics of opera, Broadway shows, and other performances.

In this reframing of the understanding of how musics are technologically mediated, add once again a possible improvisation trajectory to what seems to be recorded music. Here I now take note of transforming what is initially recorded music in a "performed" or constructed music direction:

- Perhaps the simplest and most direct transformation of recording technologies into instrumental performance ones is the "DJ" use of records being played, which are then, hands-on, manipulated by the DJ thus changing the sounds. In this case, the "same" technology which produced a music for "passive" listening, is changed into a transformed music.
- Second, and drawing from the above history of recording technologies, a more deliberate and creative transformation comes from a twentieth century example—the music of Györgi Ligeti. Here, he utilized a cut and paste, or bricollage construction of a composition is made from previously recorded sound bits to produce entirely new and different music. The end result is nothing like the previous musics which were recorded, but is a new music with its own gestalt sonic character.
- In both the previous examples, the musics mediated by recording technologies are produced by rearrangements and reconstructions.

A third transformation with deeper implications relates to the initial production of recorded musics. The emergence of the sound studio which bears a direct parallelism with a science laboratory, provides the possibility for further manipulation and transformation of sounds. The sound studio technician—like the lab technician—tunes and tweaks the sounds on the way to being recorded. My point here is that one needs to envision here the much more "corporate" or team involved in music production. It is not simply the individual or group playing which makes the music; it is rather the whole complex of processes, persons, and technologies which produce the music.

Then, as previously noted from our recording technology history, this growing aggregation of parts from which music is produced, is largely a twentieth century phenomenon with all of the above getting as far as a movement from mechanical to electronic mediating technologies. With the late twentieth-into-twenty-first century development we reach the level of digital and computer assisted sound producing technologies. This development signals a final break from the implicit "copy" or "re-produce" model of sound production and shifts to synthesized, generated sound which is no longer necessarily based upon copied or recorded sounds—digital-computerized music does not need an "original," but is itself an "original." Today, of course, there are many examples of such musics, ranging from techno to electronic synthesized to totally transformed musics mediated by computerized-digital technologies. One innovative example comes from the work of Felix Hess, a physicist-turned-artist. His book, *Light as Air*, containing of course a CD of his produced musics, takes recording into a different style of construction. Admittedly, he often draws from "natural" sounds in the sense that one piece uses sensors which use the windowpanes of apartment complexes as speaker diaphragms. The panes vibrate when sound is produced within the room and the sensors pick up this sound and "record" it. But then the sounds thus collected over possibly days, is computer tomographically time compressed so that a twenty-four hour period is reduced to eight minutes of played sound, a new music.

What I am suggesting is that the boundaries between recorded music and a new, complex, and technologically developing music production are blurred. It is here that I can return to my opening with a Foucault-like episteme. What I discern in the histories I have traced, is that musics which include instrumental technologies, may now be reconceived and understood in many ways as parallel to what has happened in science instrumentation. The old image or episteme of early modern science was one which romanticized the genius individual—Galileo, Leewenhoek— each producing his own instruments, telescopes and microscopes, and bravely discovering through newly mediated observation, the new phenomena which made early modern science what it was. This is surely not the episteme of late modern, or possibly postmodern science. Rather, the new instruments are the large, complex colliders and particle collectors, which are "played" and tuned by many scientists and technicians in the Big Science of today.

Music, in a different but often parallel fashion, and perhaps most clearly seen under my image of the large escaped canary of recording technologies, shows precisely the same kind of shift. But we have rarely recognized that instruments can be not only simple, relatively small and perhaps individually played, as with my gourd bow example, but they can also be large, complex, high tech, and communal in the production of new musics.

This, in turn, casts a different perspective upon recorded music. Much recorded music today is a sort of doubled reproduction. Recorded results of studio produced and manipulated musics, are pretty much what books are to textual productions. They are the "calcified, materialized" artifacts which we can pick up and read or reread at will and leisure. I could have begun with the book metaphor, of course, but ending with it perhaps helps with understanding some of the ambivalence so many music theorists feel about recorded musics. When reframed by a book metaphor, while there remains vestigially something of "copy" notions of an original, such "copy" and "reproduction" notions become weaker. No one expects readers of books to regret not having an original hand written manuscript instead of a nicely printed book. And such a reframing also should weaken any elitism which often gets expressed in disdain for recorded music from music theorists. In parallel, from a humanities perspective, I doubt most humanists would ever think of decrying the world of books as texts in the way in which some music theorists decry their auditory counterparts, recorded music. Neither the book, nor the record is dead. And both are only variants in the wide, wide world of language and musics.

NOTES

1. Michel Foucault, *The Order of Things: An Archeology of the Human Sciences* (New York: Vintage Books, 1973). Foucault is citing T. Camenella, *RealisPhilosophia*, 1623, p. 98.

2. I am here following the numbers of the US standard; those of the UK are on a smaller scale.

3. Mercater et. al., *Science*, Vol. 216, No. 5572, 24 May 2002, p. 1380.

4. *Science*, Vol. 291, No. 5501, 5 January 2001, p. 52–53.

5. "African Music," *Encyclopedia Britannica*. 2007. Encyclopedia Britannica Online 11, http://www. Britannica.com/eb/article.57074.

6. *Ibid.*, 57074.

7. J. D. Williams, "The Art of Music," South African Archeological Society Newsletter, 1981, p. 8.

8. Encyclopedia, op. cite. 57–74.

9. There has been a five year project located within the Free University of Berlin which has been investigating the parallel history of instruments in both science and art. My contribution to this project has appeared as, Don Ihde, "Die Kunst kommt der Wissenschaft zuvor. Oder: Provozierte de Camera obscura die Entwicklung der modnernen Wissenschaft?" Instrumente in Kunst und Wissenschaft, eds. H. Schramm, L. Schwarte, J. Lazardzig (Berlin: Walter de Gruyter, 2006), p. 417–430 (an English version is forthcoming).

10. Still following a long interest in instrumentation, a number of works are now beginning to be published. See, Don Ihde, "Models, Models Everywhere," *Simulation: Pragmatic Construction of Reality, Sociology of the Sciences Yearbook* 25, eds. J. Lenhard, G. Kuppers, T. Shinn (Dordrecht: Springer, 2006), p. 79–86.

11. Trevor Pinch and Karin Bijsterveld, "Breaches and Boundaries in the Reception of New Technologies in Music," *Technology and Culture*, 44.3 (2003), p. 538.

12. *Ibid.*, p. 54.

13. Martin Heidegger, "On the Hand," cited in Michael Heim, *Electric Language: A Philosophical Study of Word Processing* (New Haven: Yale University Press, 1999), p. 210–211.

14. Felix Hess, *Light as Air* (Heidelburg: Kehrer Verlag, 2001).

SIX

Synthesizing Sounds

Vienna, the European city with deep musical history and credentials, has often been the site of my own philosophical events concerning music. It was in Vienna, in 2000, that Trevor Pinch, my then still living composer son, Mark, and I, did a panel at the Society for the Social Studies of Science where the topic revolved around Pinch's researches on analogue synthesizers—the Buchla and the Moog. But on another occasion I did a solo presentation on various kinds of musics at the University of Vienna with what turned out to be an interesting "purist" filled audience. I was, of course, discussing the various kinds of instruments as they developed historically. During the discussion there were a number of passionate persons who strongly claimed that "the only authentic way to listen to Baroque music was with Baroque instruments!" In one sense, I had no problem since virtuoso playing on such historical instruments can certainly be an excellent musical experience—yet what struck me as a sort of "Heideggerian authenticity" also seemed somewhat ideologically reductive. As noted in the previous chapter, such a traditionalistic restriction would, in science, be destructive.

In the previous chapter I ranged into deep history—musical sound making through instruments is indeed ancient and the history and proliferation of instruments is amazing. In this chapter I want to focus upon what might be called postmodern electronic instruments—synthesizers. However, before doing this, I will make two preliminary observations. Musics obviously belong to many cultures. While in today's world there are a few music-iconoclastic cultures, ones which forbid some or even all musics (the prize winning movie *Timbuktu* illustrates this phenomenon), most cultures celebrate musics. They do so with pluricultural variations. For example, in musicological history one can note that similar to my Baroque purists, preferred sound styles can emerge which favor certain kinds of sounds over others. In the early classical periods of Europe the well-known preference for vocal sound styles within Italian culture, contrasted with the preference for instrumental styled sound in Austro-Germanic culture has been noted. More recently, I have noted how with South American, particularly Brazilian bird songs play roles in—for example—Simon and Garfunkel compositions. There is a local tradition which like the earlier Italian tradition favored human voice tonality, or Austro-German traditions favored instrumental sounds, so here bird song was another favoring of sound qualities. Also in the previous chapter I noted the antiquity of both bodily musics—singing and dancing—and instrumental musics from percussion to string to wind and other instruments. Such sounds are enigmatic: they are not "natural" in the same sense as animal songs, nor bodily-perceptual without a performance-mediational role. And while instrumental sounds are or can be analogous to natural, animal, or human bodily expressed sounds, they are also different. In premodern instruments, it could be noted that differently produced, often bodily vibrations produced the sounds: lip vibrations for brass; reed vibrations for woodwinds; breathing techniques for recorders, flutes, and the like. However, the primary focus of this chapter will be upon what could be called *postmodern sounds,* with my narrative following the development of electronically produced, or synthesized sounds. These are of late modern into postmodern instruments and as with all instrumental histories, are of several variations.

Trevor Pinch, mentioned above, plays a role here. Not only is his work on analogue synthesizers crucial in sound studies, but the noted collaboration in Vienna was followed by his participation in one of my techno-science roasts and a conference on synthesizers at Stony Brook. I have always been deeply appreciative of his insight into what might be called my postphenomenological attitude toward *music* in all its variations. In *Postphenomenology: A Critical Companion to Ihde* (2006) he says, ". . . and when it comes to music [Ihde] considers all genres: The Indian raga, Beethoven, and the Rolling Stones are all mined for examples. . . . The beauty of sound and Ihde's approach to it is that there is no 'high church.'"[1] I've always appreciated that comment; it reflects my phenomenological practices—listen to variations and do not be reductive.

So, now to synthesizers and the electronic instrumental production of sounds. The first synthesizer—an electronic musical instrument—appears to be Leon Theremin's "theremin," invented in 1920. Once again, the invention was accidental in the sense that Theremin did not set out to design a musical instrument. Instead, he was experimenting with electromagnetic fields in the hopes of making a device which could measure gas densities. His machine was an L-shaped antenna-EMS field producer which produced two differently shaped fields in the space within the L-shape. The "aha" surprise was that when he moved his hands within that set of double fields, weird, ethereal sounds were produced. One could say that these were the first synthesized sounds and Theremin and others noted that they were not like any *natural* sounds. These were instrument produced sounds, "synthetic" sounds for the vocabulary of the day.

The theremin turned out to be playable and Theremin himself learned to play Russian folk melodies on the instrument. Playing it was odd in the sense that unlike previous instruments one did not actually touch it (as one would with strings, keys, mouthpieces, membranes, etc.) but by "air playing" one could get a melody. This is a different style of embodiment, not so much through the instrument, but more like a free-form dance movement of one's hands in space. In one sense it was easy to learn. Theremin himself taught people to play it, but to become virtuoso in performance was very difficult and as either a standalone instrument or as an orchestra component not much music or many performances were given. Although some orchestras and more individual soloists produced concerts, the ethereal sounds were more often taken into several genres of early film—particularly science fiction, horror, or fantasy genres. For sound tracks the Theremin did succeed. (In chapter 9, I shall note a very different adaptation of the theremin as a surveillance instrument.)

If the theremin produced EMS wave music, other forms of synthesis were also possible. In the case of Gyorgy Ligeti (1923–2006) if the therein was the production of nonnatural sounds, much of Ligeti's electronic compositional work was a sort of experimental bricolage in which he took sound bits—sometimes from cut and paste tapes—and refashioned them to produce new sounds. His compositional career was very complex and he went through several styles of avante garde experimentation. But although many of the sound bits or atoms were cut and paste, the result didn't sound "like" any traditional music. And the connection already noted with the theremin continued with Ligeti—his music was used for the soundtrack of Stanley Kubricks' wildly popular film *2001: A Space Odyssey*, music which drew from some of the earlier "Atmospheres" composition of Ligeti. Chronologically, this style of synthesis was mid to late twentieth century.

The next phase has been most thoroughly dealt with by Frank Trocco and Trevor Pinch in *Analog Days* (2002). The analog synthesizers, Moog and Buchla, were mid-century inventions (1964–1975), both by tinkerer-inventors who, as so many at the beginning of the digital age, were kit builders often working out of garages. Pinch emphasizes two seemingly drastic differences in style: Moog decided to adapt keyboards as his playing control, while Buchla eschewed this and wanted an instrument which would break with keyboards. His control system had dials, pressure surfaces, and thus called for newly developed embodiment skills, whereas Moog could count on preskills. I do not want to miss noting in passing that Moog was familiar with the earlier theremin and in fact produced them—even today his company sells theremins or theremin kits.

But the analog synthesizers were very different from the theremin—they could be made to "sound like" anything else as Pinch points out: "There is something unique about the synthesizer. Unlike almost any other instrument it can be used to emulate or imitate other instruments . . . at the push of a button [it] can be made to sound like almost any other instrument (including other synthesizers)."[2] Commercially, it was the Moog with its preskilling keyboard which became the success. Yamaha, Casio—and later Kurzweill—sold hundreds of thousands to millions of relatively cheap synthesizers, and with a touch of the button one could have piano, koto, harpsichord, or organ ad infinitum sounds. As with most electronic technologies, originally high priced ($15,000) instruments, mass produced and miniaturized, now sell for less than $100. Our own one remaining digital piano is a Kurzweill which used the recorded sounds from a high quality Steinway for its piano sound. Here we are with a mass produced, electronic form of a Ligeti tape splice! But, if Pinch's interest, and the analog designer intent was the synthesis of any instrumental sound, that is not what emerged as the most interesting trajectory of the analog synthesizers. Synthetic sounds were electronically produced and thus they could be *shaped* through various electronic manipulations. Moog was an early user of what I will call *hybrid* synthesis. He added to his manipulation machinery, for example, an oscilloscope, as Pinch describes it: "With the aid of an oscilloscope he used his *eyes* to *see* the shape of the waveform, with the aid of a loudspeaker he used his *ears* to *hear* the sound of the waveform, and with the aid of a voltmeter, he used his *hands* to *tinker* with the circuit producing the waveform."[3] I cannot better put this as a new, multimedia form of embodiment practice. It is also an early development of both acoustic and optical imaging for sound shaping.

Synthesized sound for music, however, did not follow designer intent. What musicians took from electronically produced sound was not its ability to mimic either other instruments or other synthesizers rather, musicians became interested in the new and unique sounds of synthesizers. Pinch recognized this and cites several musician's comments: "I had no interest in using the synthesizer to create instrumental sounds . . . the sounds were never as good as the original acoustic sounds", Jon Weiss.[4] Or, again, "I wanted the Moog to be the Moog," David Borden.[5] And, as an example, "One sound I remember distinctly was a plucked string, like a bass sound. Then it would slide down—it was something I could not do on an acoustic bass or an electric bass. . . . The Moog not only sounded like an acoustic or electric bass, but it also sounded *better.*"[6] Actually, this has already become the synthesizer trajectory—the ethereality of the theremin, the tape splicing of Ligeti which didn't sound at all like bricolage, and the new sound capacities of the analog synthesizers combined with the embeddedness in emergent film-recorded styles of music, had already demonstrated that this was a new sound. And the synthesizers also were reflections of precisely the new imaging technologies which were electronic, digital, and enhanced by computer tomography. The final step for our time is the turn to the digital synthesizer.

In a transition, I return to the period when my son Mark (1986–2012) was highly active as a young composer. For a number of years he composed and won composition awards, first for piano, but then for electronic synthesizer compositions. Herb Deutsche, frequently noted by Pinch as a collaborator with Moog, was one of the award judges. I want to describe a few of the developments which occurred during this time since they lend insight into a claim I want to make about the synthesizer development. First, early in Mark's composition career, which entailed a Roland digital piano with composition programs such as Midi, what would happen is this: Mark is composing and the program can translate his playing into a printable score. But with multiple runs and playful variations, Mark discovers that he can compose for 10, 20, 30 or more fingers but for ever so many more fingers than a solo or duet piece. Or, he can accelerate the time such that no human could play as fast as the program allows. Or, he can reverse the melody lines, to variations which finally crash at the borders of the program. The human-computer interface here is reaching a different set of human capacity limits than the perceptual limits noted earlier—these are embodiment limits.

I now introduce a different set of at-home experiences. Linda, my wife, knew that I liked *simulator games,* so for a birthday she got me a state of the art flight simulator game. It came with a small set of airplane choices, an old biplane, a Cessna, a fighter jet, each with a program which took account of the capacities of speed, climb, etc., of the different airplanes. But also, you could—as with so many computer games—choose a point of view, which included a position within the cockpit, looking out the windshield, with control panel and stick or wheel before one; or one could have a distant *third person* (some called it a disembodied position) or an *elsewhere or distant position,* and control the airplane as it were a drone—out and over there. My first choice was in the cockpit "embodied" position and my first attempts were to takeoff, navigate, and try to return to my proper field. I failed and got lost over Illinois cornfields. Mark, on the other hand, chose the "disembodied" or drone-control position and joyfully flew the craft into one of Chicago's tall buildings to see the plane disintegrate as it fell to the ground. The program, in other words, builds into the game a wide set of possibilities both for the vehicle and the player. Or, for example if one is playing a warfare game, one can "kill" multiple foes—but one can also be "killed" multiple times. This may seem like a detour but the point I am now making is that synthesizers do like things in their sound production designs. Synthesizers are *like* computer games! In one sense this should be expected since their designers belong to the same digitally textured culture. The game culture permeates design in electronic culture.

I now return to composition via synthesizer. For the years of his productive compositions, Mark enrolled in a junior composition class in Stony Brook University's music department. While they had already retired the Moog and Buchla synthesizers to a sort of museum lab, the digital lab had the latest machines which by now had the standard optical-acoustic hybridity. When producing new sounds one could both hear and see via a projected visual wave form, the sound. By changing whatever parameters one chose, one could change the sound which could also include modifications not perceivable to the ordinary listener. Out of this came Mark's "Archangelmecha" (2001) composition which added to the digital synthesizer sounds from his drum synthesizer. It won a Herb Deutsche prize.

I will conclude with one set of recent collaborations: in 2010 Simon Fraser University sponsored a conference which had the theme of movie sound tracks—just right for, as noted, the closeness of much synthesized music. I was to give a keynote address and chose to do it on acoustic imaging using clips from a wide variety of electronic compositions both by musical composers and performance artists. Barry Truax, a major Canadian electroacoustic musician, participant in the World Soundscape Project, composer ("Riverrun," "Wings of Nike") with a special interest in granular synthesis (very small sample sizes), was my host for a workshop in his experimental studio. For my part, I used my collection of sometimes wild experimental CDs discussed in various chapters, which he fed into his state of the art technology which with hybrid fashion included a glowing color wave display so we could all see the "wave shapes" my clips contained. The collaboration continued in the years following and will result in a publication to which we both contribute, *The Routledge Companion to Sounding Art* (2016 forthcoming).

Synthesizer sound is malleable. One can, using opto-acoustic modeling, simply shape the sound via manipulating the sound wave. Electronic sound, as the sound of any instrument, also tends to favor certain qualities—here many users note that timbre is very important with overtones. This is an unintended side effect which oddly reverberates with that most ancient single string bows which, with skilled play, also relies on timbre and overtones for richness.

NOTES

1. Trevor Pinch, "Voices in the Electronic Music Synthesizer," in Evan Selinger *Postphenomrnology: A Critical Companion to Ihde* (SUNY Press, 2006) , p. 50.

2. *Ibid.*, p. 51.

3. *Ibid.*, p. 54.

4. *Ibid.*, p. 59.

5. *Ibid.*, p. 59.

6. *Ibid.*, p. 61.

SEVEN

Embodying Hearing Devices

In order to continue to lecture, travel, and teach in my eighties, I now must wear a pair of hearing aids, acoustic technologies. Mine are small, in-channel, and state of the art digital devices. These have three programs: one for everyday use; one tweaked for the "cocktail party" or restaurant setting in which near, ambient noise threatens to overwhelm hearing one's conversational partner(s); and one for telephone use. And, they are *expensive!*

There may be some personal irony here; more than three decades after I was engaged in all sorts of auditory experiments, which became the basis for *Listening and Voice* in its first publication I had acute hearing and definitely did not need an acoustic technology as prosthesis. But, from then to now is a long time, and I propose to follow a phenomenological itinerary along that way and analyze the process and experience of "embodying" such devices. I have described this process of embodiment in a series of previous works, beginning with *Technics and Praxis* (1979) and more definitively in *Technology and the Lifeworld* (1990). When we humans use technologies, both what the technology "is" or may be, and we, as users undergo an embodying process, we invent our technologies, but, in use, they "reinvent" us as well.

In both books mentioned, I used optical examples beginning with eyeglasses. Embodying eyeglasses, I contend, is much easier and somewhat different than embodying hearing devices. Again, I begin autobiographically but any optician would recognize the implied patterns as symptomatic rather than individual. By my late fifties, I began to notice that it had become difficult to read telephone directories and the *New York Times*. So, after the proper examinations—themselves entailing sophisticated optical devices—I was given a diagnosis and a prescription for reading glasses. These I still wear, but I do not now need eyeglasses for nonreading purposes. However, there was a temporary time when I did need such mediating devices due to two closely timed accidents: chasing a porcupine in the dark after I heard him gnawing on my Vermont cabin, I encountered a tree twig and scratched my cornea. Not long after, one of my then infant children, with a finger poke, scratched the other cornea. So, after medical treatment, and then an eye exam, I had to wear prescription glasses for about a year while my corneas gradually returned to their proper shape. So, I had to learn to see through glasses, to embody them.

If I revert to a third person, anonymous description, I might recognize that what glasses do is "correct" vision, in this case, to compensate for deformed corneal shapes. But just putting on glasses does not simply "snap" vision of the world into its now simply corrected sighting. Instead, one has to "learn" and bodily accommodate to wearing glasses. Phenomenologically, seeing is a whole body experience. There are discernible changes in depth and motile perceptions, one is aware of this in the simple act of walking. The same happens with every new prescription. It is my bodily orientation which is the nomadic part of this experience. This is even more noticeable if one wears reading glasses without taking them off to go get a drink of water—these mediating technologies produce a repeatable distorting effect which is quite perceivable. But, in my experience, and in those who have related theirs to me as well, embodying new eyeglasses to the point where they are nearly functionally "invisible" is a very quick process, maybe a day or two at most.

This is what I have previously called an *embodiment relation* with, in this case, optical technologies. I relate to my environment, my "world," by means of such technologies and if they are well functioning then experientially they are "taken into my very sense of bodily experience." My awareness of wearing glasses is a fringe awareness which gets interrupted only when there is back glare, or when the glasses slip off my nose, or when the lenses get dirty and smudged, when, in other words, something diminishes the normative transparency of the optics. "Breakdown" is a well-known phenomenon, made popular in the famous example of Heidegger's hammer—his claim was that only when something is missing or broke does the set in assignments and involvements become clear.[1]

With hearing aids, however, the technology of interest is an acoustic or auditory technology, a hearing "aid" ideally should function parallel to the visual eyeglasses example. Unfortunately, auditory transparency is much more difficult to attain, a fact well recognized by audiologists and others. A significant number of people attempt to use hearing aids, but the difficulty of embodiment is sometimes such that they give up.

I am not sure when I first became consciously aware of my slow loss of hearing. As with reading glasses, I became aware, with aging, that my hearing was not as keen as it once was. In academic life, verbal situations are of focal importance and at some point I began to be aware that the "cocktail party" hearing problem began to occur. That is, in a conference or reception setting, background conversations and other noise seemed to intrude and overwhelm my ability to hear what nearer conversants were saying. Similarly, in large lecture classes, the questions from the back of the auditorium seemed too faint or indistinct. Beginning to recognize that I was experiencing hearing loss, and remembering all those sixteen hour days of the loud noise of the two-cylinder John Deere tractor I used on my father's farm, I wondered if I had acquired "boilermaker's disease." However, I had also already noticed that many of my age peers among the faculty already wore hearing aids and their possibly loudest boyhood experiences were with stickball games. And, I knew I was too early to have the problem of "rock concert disease" either. But the clincher came via a more technological means.

While at an American Philosophical Association meeting in Boston, my wife and son were off visiting the Boston Science Museum and later invited me to revisit the place. One of the exhibits had to do with the senses, including hearing, and one could put on earphones and turn a dial to find out how many cycles per second one could perceive. I put them on and discovered that the upper range of my hearing only went up to 10,600 +/- per second! Cognizant from descriptions of "normal" hearing that humans can hear between 20–20,000 cycles per second, I was shocked. I didn't say anything, but on reaching home, quickly went to my *Macropedia* to find out what the "objective" situation was—I was relieved, in one respect, to discover that my range was relatively "normal" for someone my age (mid-sixties then). What counts as normal, apparently, is also age related. Recently I read a newspaper article in which a store owner, constantly having teenagers hang out in front of his store, wanted to prevent this pattern of behavior. Somehow he knew that the hearing capacities of those below twenty differed by a small frequency range from those in their thirties. He installed a noise device which broadcast this upper end of the frequencies heard only by those at the lower age level, but remained undetectable even to those in their thirties![2]

At this point, I decided to experiment. I purchased a single hearing aid, advertised as digital but requiring no special set of auditory exams. And, it did work for a short period of time—it amplified to the degree that I could better hear the questions in the back of the room and in low background noise situations. But its limitations were equally obvious.

Unfortunately, with more age, I experienced more hearing loss, now quite perceptible, particularly in the conversational contexts already mentioned, but also in a home setting. So, after a series of audiological tests which showed the degree of hearing loss, I ended up with my pair of digital hearing aids. Undergoing such tests, again with complex technologies tuned to various auditory phenomena, I began to learn things I never knew before. Part of what I learned was that even hi-tech, digital aids cannot restore frequencies lost, although within those which remain, these devices can be selective regarding enhancement or reduction and other manipulations. Again, this reflects my earlier claims about technologies—each transformation of experience displays an amplification/reduction structure. Eyeglasses do this and so do hearing devices. But, in the case of speech, a more subtle phenomenon arises. Vowels are temporally "longer" than consonants (and thus are more easily amplified); but since speech depends upon the patterned gestalt of both vowels and consonants, the loss of consonants complicates hearing and understanding speech patterns. Digital devices, within limits, can enhance consonants. And, I experienced this with my first set of prescribed hearing aids. These were state of the art devices, and they did make it easily possible to continue seminars, to allow much better auditory recognition of what the participants were saying. But, as technologically sophisticated as these were, once one was in the "cocktail party" situation, near conversants continued to be overwhelmed by the amplification of ambient surrounding sound.

If I refer back to *Listening and Voice,* it will be recalled that sound is simultaneously experienced as both surrounding and directional. Hearing aids, however, cannot simply match this phenomenon. And that was particularly the case with my first set. I had been advised to get a pair, and not a single device, even though I had more loss of hearing in one ear than the other. The reason was that I needed to relearn to hear directionally and this was presumably better accomplished if one began from the start with a pair of devices. Recall a parallel from optical history: monocles were once used, but are rarely, if ever, seen today. Instead, eyeglasses as a "pair" are worn, although one lens may have a different grinding prescription than the other.

At first, I have to admit, while I recognized the improvement my devices provided, particularly in the conversation settings of home and seminar, the overall experience of hearing was clearly not anything like optical transparency with eyeglasses. My audiologist confirmed that this was, in fact, the normal experience for first time users and urged me not to constantly remove the aids when I was not in the situations where they functioned best. He used the now popular "the brain must relearn the process, so it needs the constant use to do this," also translated into, it takes a long time to become accustomed to hearing aids. My acoustic technologies remained quasi-opaque, although the recognition of this opacity was phenomenologically complex. For example, if I listened to music, if I didn't know the piece, I could not be aware of what I was *not* hearing (e.g., the higher frequencies), but, if I listened to a piece which was familiar, I could quite distinctly become aware that I was missing what I remembered hearing in my listening past! Thus, familiar pieces sound "odd," compared to how they "used to" sound. Other noticeable differences apply to "tinny" sounds in some of the retained sounds, particularly when coming from a stereo in another room. The phenomenon of what is lost, compared to that simply unheard, was evidenced in a seminar which Judy Lochhead and I sometimes share on the phenomenology of music. One evening we were listening to a CD of sounds recorded by the ethnologist Stephen Feld. He had recorded the very subtle sounds of water dripping, running, etc., in the New Guinean highlands and was making a case for the auditory-dominant language of this tribe.[3] But when we played the CD, it was "silent" for me for the first minute or two—yet all the others were obviously hearing something, so, only indirectly, did I become aware that I was missing something. (Chronologically, this event preceded my use of hearing aids.)

A counter phenomenon also occurs. In *Listening and Voice* I made reference to the fact that we always hear ourselves differently from the way others hear us. Physiologically, this is because we hear via bone conduction as well as through the sound waves carried environmentally. So, the "acoustic mirror" of the tape recorder is always a surprise at first. But, hearing aids *amplify* bone conduction sounds—thus, my self-perceived voice is much *louder* to me with hearing aids than without. This complicates relearning of auditory projection when speaking, and even close up conversations. My voice to myself seems "too loud" and so I try to compensate and end up having others tell me I am speaking too softly—or, sometimes too loudly. This is the inverse of what is foreground and background in nonacoustic technology hearing.

All of this continued to be noticeable with my first set of devices, but as I approached the second anniversary of use, I lost the pair while flying from a conference in California back to New York (I take them out to listen to earphones if watching a movie or TV while in flight), and after not being able to find them or retrieve them via lost and found, I ended up getting a "new, improved" set. Indeed, in the two years of use, electronic and miniaturization technologies had already *improved upon* the previous technologies. The new set automatically "communicated" between the individual devices (if you change a setting on one, the other automatically picks up the change); there was a more refined set of programmable selectivities regarding frequency changes; and one or more programs could be made *directional* (*two* tiny microphones in each device enhance the sense of the directional, almost like having four ears!). All this for only $3,000 more than the original pair!

In certain ways, there were clearly perceptible improvements. Speech was indeed more distinct than with the previous aids, but most of all, in the second "cocktail party" setting, the directionality allowed me to hear close conversants *even though the background noise remained amplified* more than it would have without the devices. At least the cut-off point was better, but the amplification of background noise and the amplification of one's own voice still remains noticeable and frustrating. Technological

"intentionality" is simply not the same as one's ordinary bodily intentionality. At the least, one can conclude that acoustic hearing devices are
very far indeed from any of the hyped, utopian "bionic" beings of entertainment and science fiction imaginations or of virtual reality dreams.
And, hearing aid embodiment does not come with either the same ease or
degree of transparency that eyeglasses or optical technologies seem to
have. Acoustic technologies are both more complex than the relatively
simpler optical ones, but are still at a relatively early stage of development. But, I would contend, this is not because vision is in any way a
"superior" sense, or a simpler one.

Hearing, however, is highly multidimensional—it implicates balance
and motility in ways which implicate whole body experience intimately.
Indeed, I would contend that technologies which come ever closer to
being prostheses for such more complicated experience are more likely to
be clumsy and less easily amenable to embodiment transparency. Just by
way of one example, perhaps the oldest prostheses were "artificial"
limbs, and while these have made various high tech improvements, they
have never yet allowed users to regain the gaits of walking, the gracious
movement of dance, or more extremely with hands, the facilities of instrument playing which preprosthetic actual limbs allow.[4] Technological
transparency, with respect to human embodiment, remains at best a quasi-transparency.

I shall not take up here alternative hearing devices, such as cochlear
transplants, but I shall briefly look at some variations from other acoustic
technologies. First, if one goes back in technological history, earlier "hearing aids" were often hearing *horns* or ear *trumpets*. A slim tube on one end
narrowed to fit into the ear, the other end fluted out similar to the shape
of a trumpet. Such devices could mechanically channel sound and thus
amplify it. In this case, the hearing horn, precisely because it enhances
directionality and dampens ambient sound, did have a certain workabil-
ity. Similarly, the stethoscope, which today remains a sort of icon of the
medical profession, worked in a similar fashion. Indeed, during the nineteenth century there were many experimentations with a whole series of
auscultation devices, of which the stethoscope became the most successful one.

Auscultation, or the amplification of sounds from bodily interiors
when listened to by skilled and trained physicians, could detect heart
murmurs, lung congestion, and a whole series of ailments, which could
not be detected either by visual or even tactile examination. Today, such
skills are in decline and are largely displaced by newer diagnostic practices which rely upon tests, or frequently, visual imaging (CT, MRI, PET
scans, for example). When thought of as a variant upon a hearing aid, the
stethoscope, better than the hearing horn, dampens ambient sounds and
amplifies sounds from bodily interiors. The tubes which carry the sound
waves to the physician exclude ambient sounds in the examination room

and carry the amplified sounds of heart or lung processes instead.[5] Again, this is an example of a mediating technology displaying an amplification/reduction capacity. These acoustic technologies remain simply "mechanical." Today, most acoustic technologies of familiarity, have become electronic, both analog and digital.

Only briefly, these more contemporary variants include acoustic amplification capacities which go far beyond the mechanical display of acoustic phenomena. Technologies which include microphone-amplifier-speaker systems can make present sounds which cannot be heard at all without such technologies. Two years ago, I saw an interesting performance art example: in the installation, one lies down on a couch-like bed and listens to the amplified sounds of earthworms eating their way through a compost pile which is located directly below the couch. This is less romantic, but in some ways the biological equivalent to the amplification and recording of whale songs which are now part of familiar science documentary or enhanced music experience. With this style of instrumentation, however, one still hears frequencies within the normal range of hearing, thus the distant analog to hearing aids is maintained while shifted to what could be called microsound. That is, the amplification of microsound is analogous to the optical microscope in that it makes the previously unheard hearable.

Perhaps the most ubiquitous acoustic technology which relates to hearing devices, is, of course the cell, or mobile phone, particularly those models which are worn in the ear. Here the variation is not directed at micro or infrasound, but upon the mediation of *distance,* which in its phone modality is that of the voices of others. I have previously analyzed the irreal near-distance, which such communication devices display phenomenologically. Geographical or near/far distances are technologically transformed into the virtually same *near-distance.* Although I, myself, have resisted the habitual use of cell phones, even I have had the experience of calling my son to find out where he is, only to find out that he is home on the first floor while I am on the third floor in my study. His voice is as discernibly nongeographic and "near" as when the call is made from home to university. This now everyday experience of near-distance has become a familiar feature of the lifeworld.

The technological capacity to produce microsound and even to translate infrasound and to diminish geographic distance, may suggest that today's digital hearing devices may also eventually be able to follow such trajectories of development. Yet this is not to say that more radically constructed sounds could escape the constraints of all human embodiment. Rather, as with all prosthetic technologies, there will always remain "trade-offs." What are called trade-offs are precisely those interface clues to human-technology relations wherein we always remain short of "cyborgean" unity. I am sure that I am not unique or ideosyncratic if I admit that I would just as soon do without my hearing aids—just as those who wear glasses would like to do without those. But, the trade-off is precisely the strength of such prosthetic technologies and I am well aware that without my acoustic devices, I simply could not do what I now can continue to do.

NOTES

1. My own discussion of Heidegger's philosophy of technology, including the hammer analysis, may be found in *Technics and Praxis* (Reidel, 1979), pp. 103–129, and again in briefer form in *Technology and the Lifeworld* (Indiana University Press, 1990), pp. 31–34.

2. The news item I refer to was from the *Independent*, read during a trip to the UK—I did not jot down the reference.

3. Steven Feld describes an acoustic-rich language among New Guinean highlanders in *Senses of Place* (School of American Research Press, 1996), pp. 91–135.

4. An especially sensitive and phenomenologically insightful account of a high tech limb may be found in Vivian Sobchack, *Carnal Thoughts: Embodiment and Moving Image Culture* (University of California Press, 2004).

5. See Bettyann Holzmann Kevles, *Naked to the Bone* (Addison Wesley, 1997), also Finn Olesen on the stethoscope in *Postphenomenology: Critical Companion to Ihde* (SUNY Press, 2006), pp. 231–245.

EIGHT

Listening to Cancer

Although the ultimate point of this chapter is to introduce a radical and new way to detect cancer cells—acoustically—my narrative will begin by looking at some of the different ways art and science practices relate to technologies. In this case, I will focus first upon the seventeenth century, late for the art Renaissance, a beginning for early modern science. In 2006, I was a participant in an interdisciplinary conference in Berlin which was, in turn, part of a multiyear project which studied instruments in both art and science. Published results were a series of volumes edited by the PIs, Helmar Schramm, Ludger Scwarte, and Jan Lazarzdig. My contribution which arose out of my early researches in optical instruments in early modern science was "Art precedes Science: or Did the Camera Obscura Invent Modern Science?"[1] Here my tale begins by looking at the Renaissance, a period which began more than a century earlier than what usually is taken as the birth of early modern science in the seventeenth century. What often is underestimated in the history of the Renaissance is that it was a period of deep and widespread technological innovation. For much of the art practice of the times, many of these inventions were optical as previously noted: the *camera obscura,* but also the *camera lucida* and a series of lens and mirror arrangements used by artists to establish what became known as "Renaissance perspective." Readers are probably familiar with the various grid technologies as depicted by Albrecht Durer (drawings, 1525) so that the curvilinear lines of a lute could be accurately drawn. Other devices were for field use, to accurately depict a cityscape for example. Indeed, the proliferation of such optical devices in the sixteenth century was crafted by artists or artisans who were associated with the various "schools" of artists. Today, we recognize that Alberti, Leonardo, and later, Vermeer, all used the *camera* and other devices in their art practices. I often think forward from this technologically compressed period to contemporary physics with its complex of likewise optical devices. Mirrors, mirror beam splitters, manipulated photon devices, atom and photon traps such as laser tweezers, gratings, and the like—these echo the studio situations of the sixteenth century artists, but now employ expert engineering technicians collaborating with the physicists. Elaborate optics are the means by which today's quantum phenomena are displayed.

So, one simple point here is that in early modernity, the invention and employment of technologies was first associated with art practice. Later, science founders such as Galileo adapted versions of these optics—as well as inventing others—to science practices. The Galileo-invented telescopes and microscopes, along with a helioscope which was used to detect sun spots, were part of this tool kit. But what about acoustic technologies? Although not as well documented, I would argue that particularly in musical instrumentation there was an equivalent burst of invention and experimentation.

Due to my own interests in musical instruments, I have on several occasions given presentations in European instrument museums. Wandering through these museums the evolution of the various groups of instruments is telling. Begin with the strings: today's classical shapes were climaxed in the Renaissance, the sixteenth to seventeenth centuries. Both the shapes of the body and its resonator holes took place in the seventeenth century and remain standard today for classical and Baroque styled music. The museums display a wide variety of string instruments, but until the period noted, no standard number of strings was established. I have previously noted single string instruments as related to archery bows in Africa (see chapter 5); one to three stringed instruments were and remain as classical instruments in much of Asia. It is interesting to note that odd numbered stringed instruments, such as the Sanshin in China and various odd numbered stringed instruments from India contrast with the usually even numbered strings for Europe. In Europe older stringed instruments had multiple strings as per six to twelve for guitar-like instruments, and the forerunners of violins, cellos, and violas—all of which in today's classical groupings have four, earlier could vary from four to twelve. There was also experimentation with the strings for both instruments and bows—various gut strings, long hair which became dominantly horsehair for bows, and the like (the term for "fiddle" comes from "heifer gut" which were its strings). Metal strings were a late arrival, around 1900, but gut strings are often still used in concerts emphasizing instrumental purity. What is said here, however, applies to one strand of musical style and grouping. Different traditions are associated with say, later electric instruments or synthesizers as noted in previous chapters.

Before turning to a detailed case study of instrument development, I want to turn to a major historian of musical instruments, Conny Restle, another contributor to the Berlin project on instruments in art and science. Restle notes that *organon* meant in antiquity "tool" or "instrument." The later and shorter English "organ" referring to a musical instrument echoes an essay by Vitruvius (third century BC) on a Greek water organ devised by Ktesibios—a vast system of pipes using air and water to produce a musical hydraulic system.[2] She points out that by 1600 the wealthy frequently supported chamber groups and had their own instrument collections.[3] In passing, she notes that Galileo, born to a musical family, was an expert lute player, designed a fret system based upon musical instruments to measure acceleration of balls down inclined planes and tried to determine vibration rates of different tones.[4] His father was possibly the first composer of Italian opera! Michael Praetorious (1620) wrote the first classification of then contemporary instruments, which interestingly show the greatest variety for woodwinds, very few brass, and he deals last with strings, both bowed and plucked.[5] Interestingly, there is no history of changed classifications until Hector Berlioz (1843). Before leaving this survey, Restle notes that at the end of the seventeenth century (1698) Bartolomeo Cristfori invented the hammer mechanics for the pianoforte. "By doing so, he changed the musical world at a stroke."[6]

I now want to turn to a detailed case study of a set of instrumental changes for stringed instruments, in particular the violin/viola/cello/bass group which also culminated in the seventeenth century. As previously noted, musics have had an exceptionally long history. This case study looks at what broadly can be called amplification, or the power of instrumental sound to be increased in power. As noted in earlier chapters, there have been three very broad periods regarding amplification. That which goes to prehistory is clearly an acoustic-resonance set of variations which make an instrument sound louder than a bare plucked string. The Bushman twanging his hunting bow did not produce any resonance, but once in a musical group, if one attaches a gourd resonator (with some hole for the vibrations to escape), the sound is amplified. Once more, remaining with stringed instruments, there were larger and larger resonance chambers, although many remained gourds. This mode of amplification remained in place throughout the history of acoustic instruments. A second stage of amplification was, of course, electric—but this is a late modern arrival. The first electric guitar was produced in 1931, although virtually all instruments today can be electrically amplified. Indeed, while some of the earliest electric guitars had solid bodies, the 1950s saw the full trend to solid-body instruments with electric amplification. This was a sure sign that hollow-bodied sounding instruments of acoustic-resonance were changed with electric amplification. A final contemporary, refined electric amplification is electronic and is built into the construction of the instrument itself as in the example of digital synthesizers. This case study, centering in the sixteenth to eighteenth centuries, remains acoustic-resonation in form.

Nicholas Makris is by profession an acoustical scientist at MIT's acoustics lab where most of his work was directed toward marine, oceanic life. But midlife he decided to take up playing the lute and from that experience began to be curious about what the *MIT News* titled "Power efficiency in the violin."[7] The phraseology of power and efficiency is clearly anachronistic, but roughly speaking, the volume or "power efficiency" of the violin is roughly double that of the lute. Makris, first talking to lute makers, then paired up with Roman Barnas, a violin maker, and with a team launched what was to become a seven year acoustics research project. They actually examined through extant instruments, museum reports, and multiple sources, a history of stringed instruments beginning with tenth century lutes, and their near Eastern relatives, the oud, on into the so-called Golden Era of Cremonese violin making in the seventeenth to eighteenth centuries. These stringed instruments remain the most desired by virtuoso players. My wife and I have been attending events of the Manchester Music Festival for over thirty years and only recently attended a chamber event with two seventeenth century strings, an English made cello made for the Royalty, and also a Cremonese violin right before our seats.

Using the instruments of contemporary science, Makris and team did accurate measurements, employed x-ray and CT scans of the instruments to try to determine how golden era strings achieved such volume and power. Placed in the eight century long matrix chosen, what they discovered was that two parameters were the most important. The size and shape of the hole in the body of the instrument, and the thickness of the backboard or what could be thought of as the sounding board of the instrument was crucial. Early lutes had simple, round holes, ouds, their Near Eastern versions filled the hole with rosette gratings—Cremonese era instruments had elongated "F" shaped holes and what is now the classical violin shape (Figure 8.1). Close examination showed that the circular hole did not differ in power from the rosette filled hole since the air flow was associated with the circumference edge, not the center. But the later "F" shape, together with a thicker sounding back produced twice the power of the earlier designs.

The next question raised by the investigators was whether this obvious improvement in resonant amplification was by design or accident. There were no obvious texts associated with the secrets of traditional instrument making, although the slow, apprenticeship, long lasting traditions of instrument making were well recognized. The investigators decided that due to fully hand fabrication, using knives to carve the holes, there was enough variation to make the process more accidental than intentional.[8] There is much more to be said about this craft tradition—for example, it is clear that the instrument makers themselves had to have

10th C. Lute   17th C. Violin

Figure 8.1. Lute and Violin Resonator Hole Shapes

"good ears" to recognize the greater sound qualities of better instruments. Indeed, the golden era aimed for "sweet sound," the sound that today's virtuoso players wish to repeat. This, in turn, has led to the sedimentation of classical Baroque instruments regarding shape, size, and often materials as well for the last three centuries. That is the case at least within this genre of music—different traditions obtain with different genres. If one turns to electric instruments, electric guitars for example, then in the tradition of this mode of amplification there has also been development to louder and louder volumes from R and B, to Rock to Heavy Metal, etc. Of course, if one refers to different kinds of musics it is to an entire complex of human practices and technologies. The classical music referred to in this study includes its culture of listening. In the chamber setting, one sits quietly. I admit I have subtle foot movements and such, but contrast to a rock or mash concert and the audience is in bodily motion, a very different listening style. And as anthropologists point out, in some cultures all music listening entails dance as well.[9]

Although the above example comes from a contemporary acoustic analysis, my next case is fully contemporary and is yet another example of how art and science practice differ and interact. This will be the case of sonification work exemplified by Andrea Polli, currently at the University of New Mexico.

Polli is one of a number of performance artists who has developed a process called *sonification*. I have previously pointed out that in contemporary imaging processes the use of computer tomography has opened the way to a reciprocal transformation of data to image, or image to data. The preferred form of this transformation—in science praxis—is usually to *visualize* the image dimension. Thus, in solar system imaging, the shapes of Vesuvian volcanoes, even if probed by cloud penetrating radar, are displayed at the end of the process in visual form. Artists—many more than Polli—began to recognize that data could also be sonified. In part this may be due to two factors which belong intimately to art praxis: first, artists perform their practices under quite different constraints than do scientists. They could be said to be experimental for experimental sake—if we can try it, do it, why not? Secondly, while I would hold that there are degrees of playfulness amongst technologists, scientists, and artists, the higher intensity of playfulness is consonant with the "why not?" experimental culture of art. It is another case in which "art precedes" science. In this case the development of sonification processes has been largely associated with artists.

In Polli's case I will use some examples of work I am familiar with. Polli came to my attention when she was affiliated with various New York groups and was a guest of the technoscience research seminar at Stony Brook. She began her sonification work in 1999 and while clearly recognizing the data/image transfer possibility, took to sonifying some quite unusual phenomena. In one case, on a CD, she transformed solar flare eruption data into sound. One "hears" the eruptions. And, unlike a visualization with a telescope, a sonified flare will not make one blind. Another sonified CD is a complex set of sonifications of major East Coast storms. These are computer simulations based upon wind, temperature, and other variables, which are made into altitude layer models. But instead of seeing a circular hurricane cloud, Polli's sonifications—at 10,000 or 20,000, or 30,000 feet—are sounded. In a loose analogy, this is sort of turning a storm into a kind of "music," but in another sense not. As with the visualization of phenomena via EMS imaging in wave frequencies beyond our perceptual horizons, the sonified data must be assigned values. In visualizations these are called "false colors" (I prefer "relative colors") and so in sonification the artist must assign sound values. The most usual way to do this is to adapt conventions from predata/image practices such as using high notes for high frequencies, etc. In other words, there is some degree of *construction* which goes into this process. Yet there remains enough isomorphism for a pattern recognition to remain—as I shall show in the case of listening to cancer.

Polli's more recent work has included equally innovative sonifications such as those she produced in a 2007/2008 project in Antarctica where she sonified the weather pattern process being undertaken by the resident scientists there. One hears the weather patterns. Polli's purpose is not to produce scientific knowledge, but one might say it does produce a different perspective which is, of course, one outcome of creative art. Yet, her participation in high interdisciplinary groups, including today many future planning groups, brings perspective change into more immediately pragmatic and purposeful roles. Part of this, evident in her career, relates to environmental sensitivity. Her appointment is, interestingly, in art and ecology.

Before leaving sonification, I want to look briefly at a very different program. One of the great joys of doing philosophy of technology is when I travel and make presentations about acoustic technics, there will frequently be someone in the audience who has been working on precisely such projects. In 2003 in the UK, Daniel Joliffe and Jocelyn Robert presented me with their CD, "Ground Station." To listen to it seems, as if to hear a deceptively simple sound—one hears a digital piano play what sounds like hyper-minimalist music, a series of a limited number of notes reminiscent of Philip Glass or Steve Reich. The same notes circulate in an irregular pattern, quite pleasing to a minimalist fan like myself.

However, the setup to produce this sonification turns out to be highly complex. Joliffe and Robert programmed the piano to play data-turned-notes from what could be called the "wobble signals" of a geostationary satellite. The task of this particular satellite is to provide the signals used by earthbound GPS users, but what these users probably don't know is that such satellites are not really stable—they wobble. So for the signal provided to run the GPS, a wobble signal must be taken into the computation to correct for the wobble. It is this signal which powers the "music" of the digital piano.

A deeper reflection occurred as I pondered this sonification. Here is a perfect example of how there could be an acoustic-based science. Any moderately trained musician, were he or she to learn which notes sounded which position wobble, could easily know what wobble position the satellite occupied at what time. Here is an auditory gestalt recognition which will recur in listening to cancer below. Or, one could also say, here is an acoustic hermeneutics, an auditory "reading" of a sonification.

It is now time to listen to cancer. This is a tale wherein science turns to sonification. I first learned of this through a short article, "Sound Bytes" in the March 2015 issue of *The Scientific American*. It's opening:

> Composer Robert L. Alexander was sitting in front of his laptop computer about three years ago, listening to a sound file that would have put most people to sleep. It was a faint flapping, like a distant flag waving in a stiff breeze, repeated over and over, sometimes a little louder, sometimes quieter . . . forty-five minutes later . . . the flapping stopped, replaced by a sound like a wind blowing through a forest . . . "the mother of all whooshes." [10]

The file was a sonification—it is no accident that Alexander was also a graduate science student—in this case he was listening to data taken from the movements of the solar wind. Thus, echoing the art practice of turning data into sounds rather than visual images, Alexander was doing a science inquiry via sonification. "At 30, Alexander is part of a growing cadre of researchers devoted to the science of sonification: converting data that would ordinarily be displayed visually or numerically into sound." [11] In typical fashion for today's science, *Scientific American* then turns to neurology and the physiology of listening to justify why sonification is so relevant. "The ear, often more than the eye, has an exceptional ability to pick out subtle differences in a pattern; which is helpful in discovering phenomena not obvious in a visual display." [12] Citing neurologists it is noted that hearing can detect changes in sound after milliseconds, much faster than the eye. Thus, the sonification of solar wind data, volcano eruptions—and as following, cancer cell patterns—is much quicker than with vision. Even more generally, "A mammal's auditory system is faster at transmitting neural signals than most other parts of the brain. . . . [The calyx of Held] . . . can release neurotransmitters . . . 800 times a second . . . the visual pathway does not have such a speedy neural connection." [13]

While I shall assiduously avoid following any path toward otocentrism or reductionism, it is clear that sonification holds great promise for connoisseur perception scientific listeners. What this sonification is doing, now in a science praxis context, is taking the EMS data/image inversion in an acoustic direction. Returning to the original solar wind example, "The file packs 44,100 measurements into a second, a year's worth of field measurements. A year's worth of field measurements . . . which would take months to analyze by eye." [14]

As I read this analysis, I was reminded of the early days of spectroscopy when in mid nineteenth century astronomers had to learn how the spectra of stars were chemical signatures of their makeup. So with these early sonographers, there has been a gradual learning process to understand what happens with the solar wind and in medicine cancer diagnosis to which I now turn.

The traditional way diagnosis of cancer cells takes place is to do a visual analysis of prepared and stained slides. Anette Forss, a Swedish nurse and ethnographer, was for more than four years a Visiting Scholar affiliated with the technoscience research group at Stony Brook. Her long-term project dealt with the introduction of the Pap smear analysis to detect cervical cancer in Sweden. In addition to the history of the introduction, much of her analysis had to do with the training of those who examined the slides. This entailed much postphenomenology insofar as the training and development of perceptual skills needed note and analysis. Combined with the interview process from ethnomethodology, Forss produced a formidable number of studies in which one visual learns how to tell healthy from diseased cells. This style of analysis remains dominant in most cancer diagnosis.[15] The more recent development for an acoustic version of diagnosis, through a variant of sonification, has taken place in the UK. Ryan Stables, again a musician-scientist, frustrated by long waiting periods for biopsies-diagnosis-reports in the UK health system, "got the idea of transforming a visual technique of identifying cancer cells into an audio method."[16] The visual process, called Raman spectroscopy, uses laser light to illuminate cells on a slide and manipulates these cells so that the molecules vibrate with photons scattered in such a way that a molecular fingerprint emerges.[17] Stables, cognizant of the auditory ability to subtly spot patterns, sonified this process so that the listener could "hear" the different patterns and determine which cells were healthy and which cancerous. That is, he took the same data and instead of producing a visual image, produced an acoustic image. Here, parallel to the assignment of colors to the patterns, he assigned sounds. The technique turned out to be both highly successful and easily taught. Stables used 150 clinicians to examine 300 sound files and within a very short time the accuracy rate has been 90 percent successful.[18] In recent presentations I have been reporting on this acoustic diagnosis, first in the Netherlands, then at Ryerson University in Toronto, Canada. In the latter, at the end of the presentation, an enthusiastic graduate student came up and described to me his current work entails exactly such training. There are several companies working of perfecting the technique which will be both faster and less expensive than the traditional visualist approach. Stables claims that they are about a year away from using sonified spectra in doctor's offices.[19]

In the last chapter of this book, I shall return to the acoustic diagnostic technics under another perspective, because it turns out that there are several ways in which acoustic technics can sonify phenomena of interest. The sonification process here takes extant data drawn from the objects of interest, and transforms or inverts this data into sonic images. Given the capacity of human auditory perception, these images can be "read" and a determination of a result obtained. But in other instrumental complexes, also coming into development, opto-acoustic or audiovisual hybrid instruments are being used. I shall address these in the context of my concluding chapters.

NOTES

1. Don Ihde, "Art Precedes Science: or Did the Camera Obscura Invent Modern Science?" in *Instruments in Art and Science: On the Archetictonics of Cultural Boundaries in the 17th Century*, Volume Two, edited by Helmar Schramm, Ludger Schwarte, and Jan Lazardzig (Berlin: Walter de Gruyter, 2008), p. 383–393.

2. Connie Restle, "Organology: The Study of Musical Instruments in the 17th Century," *Instruments in Art and Science, op. cite,* p. 257–268.

3. *Ibid.*, p. 258.

4. *Ibid.*

5. *Ibid.*, 260–261.

6. *Ibid.*, p. 267.

7. Jennifer Chu, *MIT News* Press, Mentions, Feb. 10, 2015, p. 1.

8. *Ibid.*, p. 6

9. Steven Friedson, *Dancing Prophets: Musical Experience in Tumbuka Healing*, (Chicago: University of Chicago Press, 2012).

10. Ron Cowen, "Sound Bytes" *Scientific American*, March 2015, p. 45.

11. *Ibid.*, p. 45.

12. *Ibid.*

13. *Ibid.*, p. 46.

14. *Ibid.*

15. Anette Forss, "Cells and the (imaginary) patient: The Multistable practioner-technology-cell interface in the cytology Laboratory," *Medicine Health and Philosophy*, Vol. 15, No.3, 2012, p. 295–308.

16. *Scientific American, op. cite.* p. 47.

17. *Ibid.* p.47.

18. *Ibid.*

19. *Ibid.*

NINE

Acoustics below the Surface

Puebla, Mexico, 1996. The Society for Philosophy and Technology meets in the replica library modeled upon that of Salamanca, Spain, and if one wanders the balconies and notices the ratio of books found there, it becomes obvious that this library belongs to a much earlier time. There are rows and rows of ecclesiastic histories, but what would count for natural science could easily fit into my own study. The time of the year was The Day of the Dead, Mexico's equivalent of "All Saints Day," which meant that the streets are full of shrines with skulls, corn, and harvest items, and saints and demons—and also full of roaming archaeologists and anthropologists with whom I strike up conversations. They insist that I should visit the buried pyramid at Cholula which many claim is larger than that of Cheops in Egypt. I take up the challenge and several of us take a taxi to Cholula. Looming up over the plain is a green hill topped by a church built by the Spaniard conquerors in 1594. I later learn that there is controversy concerning the size of Cholula—Guinness Records claims it is the largest human made structure by volume; other sources dispute this. Thus, while Cholula is less than half as tall as Cheops (217 feet compared to 455 feet) its base is four times larger (480 feet for Cheops 450 *meters* for Cholula). I later learn that after a 1911 earthquake, formed stones appeared through the church's cracked basement floor—eventually after sporadic excavation, it is discovered that there was a three-stage, enormous set of pyramids—buried—under the church.

Excavation to date is partial: there are 9 kilometers of tunnels inside, one side of the pyramid is excavated with the playing fields and stepped sides now exposed. My guide stood in the middle of the court and clapped his hands. An echo came back, not sounding at all like a clap, but like a strange bird call which I recognized as that of the Quetzal bird, a

sacred symbol of the ancients. My discovery was confirmed by an anthropologist who made the same identification a few years later.

At this point in the narrative I want to introduce a discussion on what
might be called "maxi-waves" or "shakes." It was an earthquake which
accidentally revealed the pyramid—and humans far back into antiquity
were familiar with earthquakes—and associated tsunamis, volcanic eruptions, and as we move into modernity, tectonic plate movements and
underground (nuclear) explosions. So in what, once again, will be a continuum, maxi-waves elicited ancient responses. Although very ancient
antiquity thought such maxi-waves were results of divine wrath or
supernatural causes, one of our first preclassic philosophers, Thales of
Miletus (640–583 BC) began to attribute earthquakes to natural causes. He
thought the earth floated on water and periodic superstorms would
cause the shaking. Thales's texts, however, are limited to fragments and
secondary interpretations. But if we now call *seismology* the science of
maxi-waves, then Zhang Feng during the Han dynasty (132 CE) was the
first to invent an instrument which could detect the direction of an epi-
center. Zhang was a polymath who was expert in astronomy, geology,
and in this case seismology. Earthquakes, of course, played a major role
in all ancient civilizations—the competing high cultures of Greece and
Persia both saw their capitals of the Acropolis and Persepolis destroyed
by earthquakes. Crumbled structures in South India, Ziggurats in Baby-
lon, all suffered earthquake damage.

Here I note that such maxi-waves were dramatic and came into human experience in stark and external fashion. Or, in contrast, there was
nothing EMS-like to maxi-waves. And this remained the case through
very much of history. The Lisbon earthquake of 1755, however, was different. Ironically, for my context here, this was a 8.5–9 Richter Scale earthquake which occurred on "All Saints Day," November 1, 1755. The earthquake, with a tsunami and fires, following the destruction of virtually all
of Lisbon and surrounding areas, was one of the strongest in human
history. Death estimates rose as high as one hundred thousand victims.
Susan Neiman, a historian of modern philosophy, published a striking
book on this event, *Evil in Modern Thought* (2002). Her thesis is that while
some philosophers used this event to argue for a benevolent God in spite
of the evil wrought. But on the other hand, other philosophers used the
same event to argue for an areligious notion of evil, including, of course,
Friedrich Nietzsche and his "God is Dead" thesis. In terms of cultural
history, then, Neiman argues that a turning point to secular ideas of evil
took place.[1]

I now return to the Puebla event, since in addition to the earthquake
as revealer of a buried pyramid, it was also the occasion for a second
group of subterranean acoustic technics. On return to Puebla, I was puz-
zled about why such a massive pyramid should have been buried, so I
sought and found a group of local anthropologists and archaeologists

and one question I asked was, "If there is one buried pyramid, there must be more?" To this the enthusiastic reply was "Yes" with a claim that only in recent years air magnetometer surveys had been made of Central Mexico and showed some 8,600 buried archaeological structures. Few had been excavated for fear of site robberies, but the results had been scientifically mapped for future reference. Here we reach acoustic technologies which, together with similar ground penetrating radar, probes subsurfaces with pulsed acoustic waves. There is here a history which again follows that traced in chapters 2 and 3. Carl Gauss invented detection instruments in the nineteenth century (1833) which detected anomalies and variations in the magnetic field. These are at best analogous to acoustics—magnetic anomalies and variations are sensory-like detections which take wave shapes. But then in the twentieth century from World War II and the Cold War sonar variants to detect submarines were perfected and these are more clearly acoustic. Geological use was also prominent to detect ore and oil deposits. Some of these devices can penetrate down 12 meters, others using explosives for pulse production, much farther as in contemporary "fracking" processes. The results for ancient archaeology are immense: structures such as temples, pyramids, cities, irrigation, and agricultural structures have been discovered and mapped in Meso-America, Southeast Asia, Northeast North America, China, and elsewhere.

Such acoustic technologies enhance and mediate the auditory capacity of sounds to penetrate surfaces here on a powerful scale. Ultrasound medical technics, such as sonography have already been noted. Bodily interiors are imaged. These magnetometer and ground penetrating radars I would call "middle sized" acoustic technics—and they have counterparts for water and fluids as well. Sonars, and other acoustic mapping technologies were highly developed during the Cold War and continue to be employed in the massive task of underwater mapping. Parallel to the night vision/thermal imaging relation previously discussed, magnetometers and electromagnetic induction instruments detect magnetic lines, distortions, and the like, whereas radar uses pulses to penetrate surfaces. Both are EMS effect technologies. Uses are highly interdisciplinary including obvious military uses (submarine mapping, mine detection, etc.), archaeology (graves, cities, solstice circles, roads, etc.) all using EMS related imaging processes with pulses.

Seismology instruments which record patterns upon squiggle graphs are to be found worldwide. Indeed, virtually any geology or earth and space science department in a large university will have a working seismometer.

Other detection devices, ranging from side scanning sonar to airborne penetrating radar, today yield discoveries with high regularity. Benedict Arnold's last gunboat, an early fully intact gondola, sunken ships, and galleons are regularly detected with sonar. Lost temples such as Maherdraparvata in Cambodia and the Sunken City in Warangal, India, pop up with regularity. The massive flooding of the Black Sea by the Mediterranean Sea some 10,000 BP turns out to have flooded an entire agricultural area, which in turn stimulated migrations into Europe of an agricultural peoples which shifted lifestyles for virtually all of Europe at that time.

But I now want to follow my continuum frame down into acoustic technologies at microwave levels. I shall take my examples from an at first dominant acoustic surveillance set of technologies. In chapters 1 and 2, I discussed early isomorphic imaging technologies such as the telephone, recording devices, and radio. In each case, EMS-like wave-like formations played important roles. Bell's voice transmissions transformed voice waves into fluid ones (mercury) and then back into vocal sound waves at the receiver. Edicon's needle technologies did the same by inscribing a series of indentations onto the tinfoil, later baked wax cylinders. These were acoustic wave transformations which took related, but distinct shapes in the history of spy surveillance technologies.

I return here to Leon Theremin's first electronic musical instrument, the theremin (see chapter 6). Although first it was a musical instrument, it's electronic receptive capacities were found to be able to collect vocal sounds as well as produce them. As noted earlier, he found the electro-magnetic fields which were projected in the L-shaped space of the antennae-generators. When he moved his hands within them, it produced weird sounds unlike any known from nature. Only later did he discover that the device could also receive sounds. In short, this was clearly an EMS technology, one which could both receive and send. Theremin was a devoted Stalin era Communist—I have seen a documentary of him teaching Lenin how to play a Russian folksong on a theremin. But in this context in which I am introducing surveillance technologies as examples of acoustic capacities, I want to point up this secondary use of the theremin.

The musical instrument, which he invented in 1920, was "remixed" into a bugging device which he placed into a large wooden carving of the Great Seal of the United States and this was presented to the new US Ambassador to Russia, Averill Harriman (1945). For seven years it hung in his office with its spy function of recording and sending conversations unknown. Early surveillance was almost entirely acoustic. Each warring or suspicious side would "bug" the other with some sort of recording device—the theremin as one early USSR instrument for the US embassy. Albert Glinsky traces this history in his *Theremin: Ether Music and Espionage* (2000).[2]

As the espionage bugging wars continued, recording devices changed—there was more and more miniaturization; ever more clever ways of hiding devices—in telephones, light fixtures, within the architecture from design on. It is a fascinating history with whole embassies sometimes never occupied because a host country found the bugs before it opened. But with the Cold War in the last mid-century, new "hybrid" technologies began to appear—although forerunners had already been partially developed by Theremin. I refer to his early use of infrared "microphones" prior to the laser.

Although a once virtually universal music technology was the system of CDs and CD players, now endangered by other technologies, the technically informed will know that what "reads" the disc in the player is a very small laser. Analogous to Edison's needles, the laser beam reads the code imprinted on the disc. The product is auditory, the music we hear, but the process is a microprocess of ever finer physical codes which can be read. So, here, echoing Theremin, but taking the laser reader into different territory, is my second tale of espionage. Unable to physically bug the Korean CIA (KCIA), but desiring to hear the conversations, imaginative techie spies realize that windowpanes can be considered to be substitute speaker diaphragms. That is, voices inside a room produce vibrating waves through the windows, such that if "read" by a sensitive enough device—here a laser reader—one can reconstruct the conversation. The KCIA is bugged. What this illustrates is that ever finer microprocesses, frequencies, miniaturization, etc., form trajectories for new surveillance technologies. Indeed, today is is possible to do readings from architectural parts themselves. Surveillance processes are ever finer readings and manipulations of EMS microwave capacities. Once in place, the technologies "suggest" trajectories of development. In early optics magnification plus resolution was such a trajectory. Acoustically, the trajectory to ever more precise readings of microscopic waves and frequencies is the counterpart trajectory.

Contemporarily, a "dark" side of the data/image inversion of imaging technologies is very much with us. The massive surveillance of our own NSA which scoops up virtually all communications data, both textual and acoustic (telephone) imaging, and other big data spying revealed by Julian Asange and Wikileaks and Edward Snowden are known and a matter of major debate.

I want to conclude with a contemporary, related, and speculative set of examples from the arts. These are examples of what I call *material hermeneutics* or mediated "reading" practices which get the things to "speak." Contemporary exhibits sometimes include analytic displays which show how artists have returned again and again to a painting, for example, and changed it. Since moving to Manhattan I have seen a number of these particularly relating to Maisse and Picasso. The analytic displays show multiple-instrument displays which include x-ray, or very low angle color shadow effects, or other quasi-transparency imaging. In the case of Matisse's "Bathers" the exhibit shows that he returned and modified this work for thirty years, and the display shows the series of earlier changes. Similar analyses of Picasso paintings have been done as well. These examples, however, are not restricted to acoustic imaging, but there have been some odd attempts at acoustic recoveries. For example, one step in acoustic imaging is the transformation of sound waves into wave forms in material recording. So, one trial—admittedly extreme—has been to realize that drying paint might well preserve sound waves occurring at drying time. The imaginative conceit was: Could a painter, talking to his model, leave an acoustic trace? As one might surmise, nothing like "Watson, come here . . ." has yet been found, but the opening to sound preservation is suggestive.

In something of a postscript, return to the maxi-wave phenomena which opened the subsurface continuum of waves. Today, the persistence of megatechnologies emerges with the new exploratory wave probes associated with *fracking,* the geological mode of extracting gas and oil from shale deposits deep in the earth. This process is not acoustic *per se,* but once underway part of the process is the injection of various chemical mixes into the "fracked" rock tables involved and one result turns out to be mini-earthquakes which multiply in the areas under use. In Oklahoma, prior to fracking there were two to three earthquakes over a .3 Richter scale per year; this year the Environmental Protection Agency counted 585 such earthquakes. Today's technoscience retains much of leftover industrial gigantism which emerged in the Industrial Revolution of the nineteenth century (see Are We Posthuman?). This land based process parallels the infrasound maxi-waves of submarine and marine acoustic probing which impacts upon particularly marine mammal life in yet another excess process, both of which leave in place Heidegger's condemnation of this style of technology.

NOTES

1. Susan Neiman, *Evil in Modern Thought* (Princeton: Princeton University Press, 2002).

2. Albert Glansky, *Theremin: Ether Music and Expionage* (Chacago: University of Illinois Press, 2000).

TEN

Multimedia-Multitasking-Multistability

PART ONE: POPULAR, OR UBIQUITOUS MEDIA TECHNOLOGIES

In this, the penultimate chapter, I want to look at what may well be a technological trajectory for today's sensory mediating technologies. I will begin with an examination of the two most common and widespread technologies in the twenty-first century: *television* and *mobile* or *cell phones*. Estimates show that, worldwide, approximately 79 percent of households have television access and social scientists hold that 95 percent of the world's population has access to cell phones. If this is the case, this means that more people experience these media than any other. And with few exceptions, televisions and cell phones are, in media terms, "audiovisual," or in science context terms, "opto-acoustic." Or, one may say that such sensory communications are multimedia. But what does this mean? What does this imply for our perceptual life?

I will also include, alongside *cinema,* which today is also audiovisual or multimedia, although its early history had different variants as "silent" or pretalkie movies, and monosensory earbud music devices such as Walkman, iPod, and MP3 players. All these technologies are "popular" technologies and do not ordinarily serve science or STS constrained purposes. Rather, they are used by millions of people in daily lifeworld contexts and thus provide insight into how these forms of mediation affect humans in human-technology relations.

Postphenomenologically, all our perceptual-bodily experience of the world is "whole body." This was the theme of my earlier *Bodies in Technology* (2002) book. And, postphenomenologically, I am reluctant to use the term multimedia, since that conjures up an old and anachronistic notion of clearly divided senses. While it is the case that media may be "mono-sensed" whole body experience cannot. I would contend that only with great effort can we bring a "single sense" into focus and even then there is no phenomenological full success. If I concentrate upon my visual experience, my hearing does not shut down nor does my sense of an upright kinesthetic stance change. Thus, a multimedia presentation is enigmatic for embodied beings. I shall undertake a variational analysis to point up some of the implications of multimedia for whole body perceivers. To begin, we also need note that AV or opto-acoustic "multimedia" are still far short of whole body experience. Taste, smell, and above all a tactile-kinesthetic bodily sense are missing; the AV/OA mediation is a *reduced* perception or style of embodiment. These mediational forms, however, are so pervasive that they are taken for granted.

First, take some account of why these two audiovisual technologies are so pervasive and are so paradigmatic for our times. Cell phones are perhaps the single most ubiquitous technology ever used by humans. They are a "leapfrog" technology in that their use spans the full range of all industrial countries—but also they can be found in rural, desert, forest areas. Last April while visiting Brazil, even in the temperate rainforest, we found our cell phone use better than that which we have in rural Vermont, our summer home. A major reason for this is the type of infra-structure required for wireless cell phones. Lacking are the roads, the wired roadsides, and what amounts to the late nineteenth century indus-trial technologies of the earlier mentioned "Barbed Wire" technologies of rail, telegraph lines, barbed wire. Here there is the cell tower, the wireless connection to geo-stationary satellites, the wireless cell phone, leapfrog-ging the world electronically. The instrument today most usually has both audio (phone) and visual (screen) capacities—plus, if smart, much, much more. Postphenomenological researchers have taken account of much of this in its socio-political role as well—for example Evan Seling-er's work on Grameen Bank "phone ladies"[1] and Galit Wellner's *A Post-phenomenological Inquiry of Cell Phones* (2015).[2]

Television's infrastructure is a bit more complex than that of the cell phone. It may be hooked into a satellite system, but also in many areas is cable connected. For the most part, it is not wireless and autonomously mobile. The recent hybridization of "streaming" with interface with the Internet brings it closer to cell phone networking. I wonder if this ac-counts for the slightly less ubiquity of television versus cell phone use? In "tech circles" there is often talk about "one big machine" and "one little machine." For popular technologies, the television today is large, flat screen, multi-connected, and the smart phone, even more multitasked; these are the big and small machines of AV/OA choice for many, prob-ably most.

Given that the theme for an acoustic technics is focally the develop-ment of acoustical technologies, the question arises for AV/OA technolo-gies how the interrelation with embodied human perceivers works. To get some distance, I will first do variations on an older, related medium— *cinema*. Historian fans of cinema will be aware that AV movies, the "talkies," began to appear in the late 1920s, the same era in which the radio became widespread. Moving image devices go back into the nineteenth century—practical still photography, recall, began in 1839. Later, Eadweard Muybridge, with his famous horse pictures with four feet off the ground (1878) and Thomas Eakins with similar studies (1879), both of whom corresponded with each other, began to produce motion studies and did early experiments in motion photography.[3] Thomas Edison, previously noted for his work on phonography, also invented a motion picture machine, the kinetescope (1891). Competitors in France, Russia, and elsewhere also produced such devices. Indeed earlier, there were a number of hand cranked machines which showed series of stills to produce motion during mid nineteenth century times, but the first motion pictures began to be produced—Paris and California competing—in the late century. The Lumiere family in Paris produced films in 1894–1895 and film studios popped up in California by 1897. Early films, silent or pretalkies came first, with *The Jazz Singer* (1927), the first fully synchronized talkie. The famous nickelodeon machine, which showed a short film for a nickel was in Pittsburgh (1905) and a version is to be found in our Weston, Vermont, country store today.

What was it like to see such pre AV cinema? First, technically, and not long after invention, films were not, in fact, silent. Music, minimally a piano, accompanied the film. The music was coordinated with the film— fast when action was fast; sinister when melodramatic; ascending, descending according to context. Music accompaniment outlasted the transition to talkies, as musical scores remain a part of cinema experience. And the pretalkies included texts, short dialogue, or description on the screen. So in this minimalist sense, cinema was pretty much from its beginnings audio-visual. Its AV, however, remained somewhat disjunctive between the sonic and the visual. The "talkies" from the late 1920s on took on a more stage-like presentation with actors speaking their lines. But the music also remained with a movie score. There are interesting historical notes here about silent actors, who while possibly good at melodramatic facial and other gesticulatory skills, had weak voices and some failed to make the transition to the talkies.

I do not know the rationale for adding audio to early silent movies— although as I have noted from *Listening and Voice,* sound fills and makes lively what with silence would be empty and lifeless.[4] With the addition of voice, cinema becomes stage-like and many early movies are very much patterned after theater, including sets and costumes. The contrasting outdoor spectaculars add voluminous sound—scores, battle sounds, and the like. Scores, too, are interesting remainders from silent films and are taken for granted, with as noted, synthesized sound scores such as those made by therimin or Ligeti style tape splicing, associated with certain film genres, or player piano with the "Keystone Cops."

Remaining with the cinema versions of AV, early synchronization posed delicate technical issues, thus *The Jazz Singer* was a technical breakthrough. But the synthesis of audio sounds with video film did not seem to be problematic. Clearly, our whole body perception is multidimensional and, except in pathological cases, "synthesized." Today, virtual reality variations do weird things with this normal synthesis. For example, in Umea, Sweden, a "reality helmet" turned sounds into visual forms and sights into sonic forms thus disrupting any normal synthesis. Think, too, of attending a classical chamber concert. The players are in lively motion, with degrees of individual motile styles and thus to listen-watch this scene is engaging. In my case, with interest in digital composition, to attend a concert which a few speakers, digital equipment, wires, is simply not the same experience! When "jazzed up" with a large screen, live players as in the Glass or Reich concerts—"Hindenberg" or "Einstein on the Beach" at the Brooklyn Academy of Music—and one is back to AV drama. Yet, minimally, one point is that we already have synthesized both the audio and visual dimensions of these presentations. We do not have to "add the parts" to experience a holistic sensory gestalt.

In the cases of the technologies above, all are AV/OA. I want to take a quick detour to an acoustic technology which fits into the popular group of technologies, but which has a very different effect upon "synthesized" experience: earbud music technologies. The first truly popular device was the 1979 SONY Walkman which eventually sold some four hundred million sets until production stopped in 2010, superseded today by iPods and MP3 players and streaming devices, including apps for cell phones. Unlike the AV/OA technologies, these devices produce auditory outputs only and they create quite a different effect with respect to the reflections of audio-visual experience mediated by the other technologies of concern here. (Ironically, the technology could be thought to be AV/ OA, since a miniature laser device is used to "read" the disc!) Designer intent could be said to be to provide music to an individual listener wherever he or she might be. And, as an earbud device, such listening is radically different from the public "boombox" music so disdained in urban settings, a technology which earbud technologies largely replaced.

I now introduce earbud music devices as a different variant. The listener hears music—one early motivator for the Walkman design wanted a means to listen to opera without having to carry a bulky cassette player. But today, as any observation would show, earbud listeners are everywhere—walking, on commuter trains, in restaurants, again ubiquitous technics. But unlike cinema, television, or cell phones, the earbud technology has a very different experiential presence. Here the auditory dimension, which can be quite dramatic and often with high volume, *does not synthesize or coordinate with our other sensory dimensions.* In terms of a sensory gestalt, earbud listening is more *disjunctive.* It "calls attention" to itself. The listener gets "into" the music. One could say this phenomenon is one whereby the mediation is more an *insulation* (although not a total separation from) our ordinary whole body perception. One safety concern has been that earbud listeners, walking or worse, driving, are too insulated from the environmental sounds of traffic and people to avoid mishaps. (To anticipate the current safety concerns over driving and cell phones, either acoustically or texting, is an AV/OA parallel.) The point here is that unlike a piano playing to coordinate or synthesize with the cinema scene, the earbud experience can be anything from insulation to distraction, to even greater absorption with the auditory. As a variant, one could, of course, simply use solid earplugs to shut out or dampen the sounds—as when I used sound suppressors when chain sawing—and this would have an insulation effect as well. But this means active listening to music is always more than insulation—it is a substitute sound dimension. Stacy Irwin has a forthcoming book in this series which examines earbud technologies in detail. She points out that one variant with users is to actually turn off what she calls "private sound" and simply wear the earbuds to socially insulate the wearer.[5] It is taken from its usual spatial listening constraint—a concert or sitting audio context—and goes with one in other activities: walking, driving, riding. There remains, thus, some kind of *contrast* with the sonic environment into which the bodily carried earbud technology is taken. These technologies—experience-wise—bring an auditory dimension to the user which, while changing the human-technology interrelation, does not belong in the same way to AV/OA experience. Technically, earbud technology is not multimedia, but it changes an entire environment and is thus an important variant in this context.

In the *phenomenology of technics* which remains as an interrelational framework in postphenomenology, once there is a mediating transformation anywhere within materially sensitive intentionality, the entire interrelationship changes:

Human—technology—World

Here (Human—earbud—environmental world) the earbud presented musical auditory presence, particularly played with volume and dramatic presence, clearly changes how World is experienced. At least it is sonically a musical surrounding; a "private sound." It is a heightened focal sound, to-the-ear, not to either bone conduction or bodily cavities. The earbud technology is a different variant, although it fits well into the ubiquitous type of popular technics. I now want to return to the AV/OA popular technologies for a different set of variations. I return to television. In this case, I am concerned with contemporary multimedia and multitasking televisions. Earlier TV was considerably more limited and served as a sort of AV version of radio, become radio with a visual display screen. (One of my favorite Gary Larson cartoons shows a family on a sofa, seated and looking at a blank wall, with the caption, "Before Television.") Today's TV, frequently linked to the Internet for various kinds of "streaming," will likely have from one hundred to two hundred channels, connections to various movie, sports, documentary, and other features adding to many hundreds of choices for viewing and listening. I open with a few of our own favorites: news—as noted in previous books, today's TV is *pluricultural*. News is worldwide with hotspots varying all over the globe. Terrorism can occur in any country; natural disasters are immediately broadcast; royal weddings and births occupy admirer's attention; scientific discoveries such as a Pluto flyby are present, *ad infinitum*. This range of display is a temporal condensation, a "now" which is also pluralistic, but which also displays a "near distance" or cyberspace character as if all is "here." The living room has more pluriculture every evening than had any medieval king in his castle. The media "worlds" are diverse and rich, but viewers are not bodily immersed in such "worlds." Phenomenologically speaking, spacetime has become multistable, it is both everywhere and here, plastic and flexible.

Cinema, another big machine, is even more dramatic. Here one of our own favorites is a simulcast, 3-D broadcast of the Metropolitan Opera broadcasts. The camera, following the point of view possibilities takes us, mediawise, right onto the stage with the actors. Enhanced sound systems are such that even for me with hearing aids, nothing is missed. Our "seats" are better (and much cheaper) than any in the opera house. This same positional plasticity applies to sports broadcasts. The viewer is close to the pitcher, or if it is Olympian figure skating the viewer is on the ice with the dancers. Once again, spacetime, multistability, plasticity, is evidenced. (Technofantasizers make even more about this with claims of the superiority of virtual reality or enhanced reality over "real" reality.) The viewer-listener shifts virtual positions "magically" as it were. Of course missing is the whole bodily sense of kinesthetic, tactile *place.* For in spite of many attempts to introduce this dimension into theater or cinema, none have come close to success, and smell and taste have proven impossible to deal with (to introduce a smell or taste, then rapidly shut it off as parallel to the screen scene, is too technically difficult to control). So, even AV/OA remains a reduced mediation. I well recall my first IMAX 3-D cinema experiences. When some object, animal, or whatever zooms out at me in my seated position, if I put my hand in its path, nothing is felt, it reduces to AV "illusion." In another variant are the types of Wii games which are interactive. The players must be within a predesignated space and various balls are virtually "thrown" at the players who must kick them back. This calls for active bodily movement, but one does not "feel" the balls, although the player can see on the projection that he or she has hit it.

It is now time to turn to today's most ubiquitous AV/OA technology, the cell phone. As with earbud devices, the cell phone, now likely a smartphone, is everywhere. When taking a New York City bus, I frequently do a count which shows that roughly two-thirds of the passengers are holding or using a cell phone. Cinema, theaters, some restaurants, caution all users to shut off the devices. While texting might avoid talk-sounds, the light-up features are obvious and distracting. When the shutdown announcement is made on the screen, the myriad "firefly" cell phone lights can be seen to blink out. Many forms of public transport have established "quiet cars" where such devices cannot be used. But the dominant scene is in urban, rural, and even isolated areas one in which a user may be seen holding a cell phone to the ear, alternating between listening or handheld texting. And, the old habit of raising one's voice while on a phone—a hangover from thinking the listener is *distant*— remains for many cell phone users.

None of this is what makes the cell phone the dominant, ubiquitous technology that it is. As Galit Wellner points out in her *A Postphenomenological Inquiry of Cell Phones,* which in part is a history of the development of mobile technology, what makes today's cell phone a crucial technology is its role as a multimedia, multitasking—in effect a complex electronic "Swiss Army Knife." It is clearly AV/OA with audio sending and receiving, a visual display screen, but to which one now adds full access to the Internet and multiple apps for multiple uses. Early mobile phones were mostly phones, at first bulky, then small, highly portable acoustic devices which could send and receive messages, although from the beginning multitasking was sometimes included. Today's smartphones have high resolution cameras, built in GPS, Internet connections, voice-texting-music capacities through iTunes, photo storage, calendar, and calculator functions and much more.

A year ago spring, my wife and I attended a large contemporary art show on Randall's Island just upriver from our apartment. I found the most interesting object, a sort of ball-shaped sculpture of probably at least a hundred hammers, all different in design. It could have been an *avant garde* icon for "Heidegger's hammer," but lacking the $50,000 for the asking price, I had to forego this art work. However, compared to the multitasking of the cell phone, the hundred hammers seem simple!

Given the complexity of the cell phone, it is more difficult to describe its general phenomenological features. However, with respect to space-time it does display the same plasticity and multistability found in the other popular technologies. For example, in telephone voice mode, assuming the connections are good, spatial perception displays the same near-distance or cyberspace presence as other electronic media. Someone on the West Coast sounds as near/far as someone in Brooklyn, or even beyond to international calls. And all the individual voice patterns, and thus the ease of recognizing known callers remains constant. Time, which cannot be separated phenomenologically from spatiality, is likewise compressed to near-now in the conversation. This is not to say that much greater magnitudes of spacetime disappear. Remote sensing, such as with Mars Explorer, displays a spacetime delay which is experienced by the controllers, in this case distance and time is vast, of course. But, the smart versions of the cell phone are also storage devices. Thus, everything from a grocery list to a photo album can be brought up on screen, and as with all photography, there can be a sense of pastness made present. Many users speak of their devices as a kind of "memory." One interesting contemporary trajectory for this visual capacity is the "selfie" in which the user images himself or herself in setting after setting. This is but one manifestation of display and exhibitionism also associated with a cell phone culture. I have not chosen here to do laptops, iPads, tablets, all of which are related midi-sized multitasking technologies as well. In all these cases, and many more, the contemporary trajectory is to multimedia and multitasking. In earlier work, I have shown how the history of writing technologies, with word processing today, has led to precisely this same malleability of practices. The visual display screen for all these technologies can appear as an opaque background for foregrounded, printed text and thus effectively two-dimensional; the same applies to screen reading; technologies such as Kindles and like devices; or, in game mode, the moving figures in the game field are projected into a three-dimensional cyberspace mode. The figures are in a space *through* and beyond the screen surface. And the same keys used for typing can then become control buttons (or replaced with a joystick or other more tactile device). In all of this there is multistability and plasticity. The screen, the keys, are not nor are only simply one thing, they fit into multistable contexts which change according to contexts.

Ironically, the technological tool, which to my knowledge was the longest lasting technology of human time, the Acheulean hand axe (Figure 10.1), displays a simple but clearly multitasking, multistability.

This tear-shaped stone befacial "axe" was probably used for scapping (hides or meat), cutting (vegetation or bones), digging, one anthropologist speculates was thrown like a discus for a weapon, and recent speculations are that highly refined and larger-than-usual versions were more like art objects to demonstrate the maker's skill, perhaps according to today's evolutionary theorist, as a display to attract sexual attention. These axes were used from 1.8 million years ago until roughly 400,000 years ago, the longest continuous use for a technology in my knowledge. It was appropriate to the lifestyle of hunter-gatherers, but lacked any of the near/far spacetime connections of cell phone technology. Multitasking, multistability has always been a potential for technologies. Today's are more concentrated and multiplied.

In terms of a postphenomenoloigcal analysis, one needs to look at all of the facets of the Human-technology-World interrelation to have even a

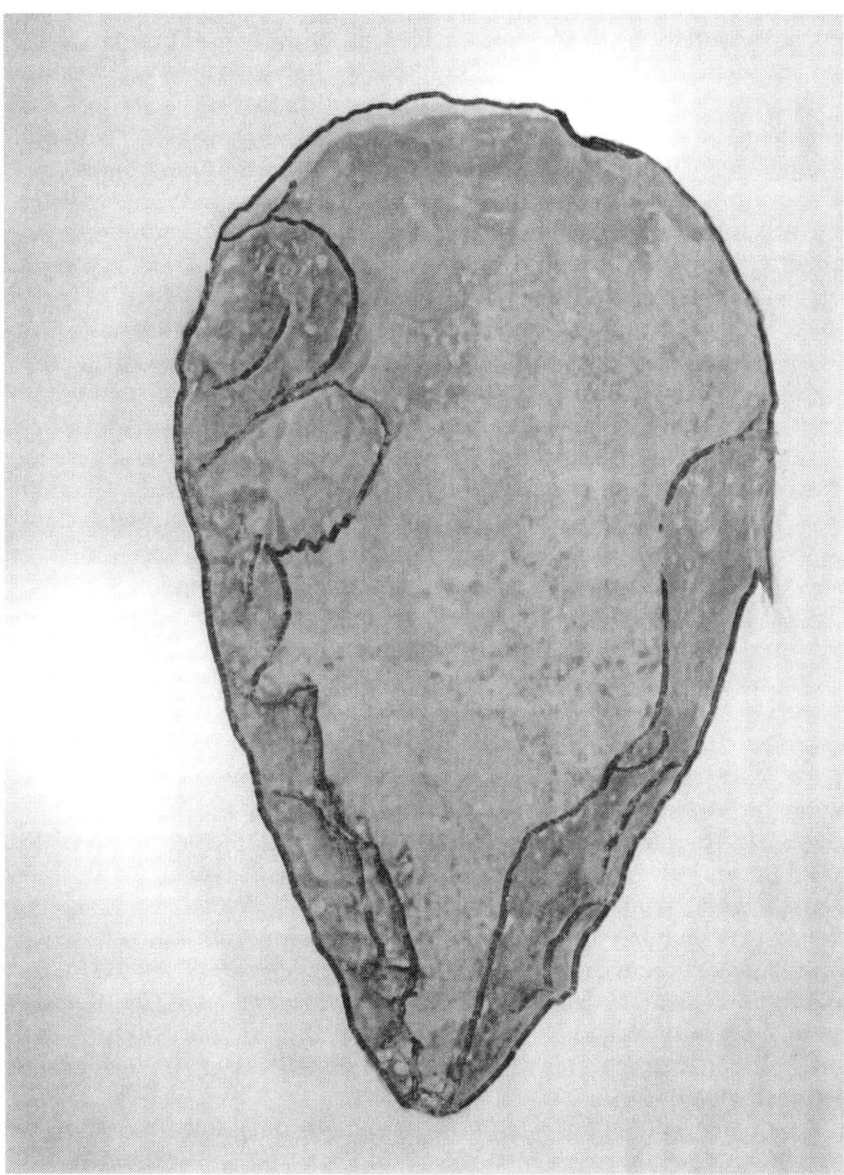

Figure 10.1. Acheulean Hand Axe

glimpse of the transformations in involved. In briefest form I will outline a few of what I take to be salient features of the contemporary lifeworld as related to our, here, popular technologies. The first feature which characterizes our lifeworld, is that it is *flat*. I am using this term which comes both from current cosmology and from Thomas Friedman's book on globalization titled *The World is Flat* (2005). Both these uses are somewhat esoteric. In the first case, "Big Bang" theory has spacetime coming into being with the explosion of a very small, concentrated entity and within a fragment of a second the universe expanded to enormous size. "Flatness" refers to our limited observability and to the uniformity, and familiarity, of the observable universe. In my periodic luncheons with Deane Peterson, my astronomy colleague, I would always ask him if the universe was still flat, to which he would assure me it was. In the Friedman book, which is about globalization, flatness refers to the way in which, whether one is in Delhi, New York, or Dubai, the same corporations are uniformly there—KFC, McDonalds, SONY, Toyota, et al.—such that the global commercial world is familiar everywhere. It is this familiarity which bespeaks what is called flat. I shall take a homier route: in the days of colonization of North America, the indigenous peoples could not understand why the colonizers thought the countryside was "wild" or a "wilderness." For our local Vermont Abenaki tribe, part of the Iroquois Federation whose governance system was partly adapted into the later Federal system of the American Constitution, the forests, streams, and mountains were simply "home." Interestingly, once I acquired my Weston property, located in the forest by a trout brook in mountainous terrain, my then very young children grew up each summer without electricity, TV, and we cooked by outdoor fire. The children often lived in pup tents off from the cabin which was our first shelter. They, too, thought of this as "home" not as wilderness. Their world was also "flat," familiar. The frogs and garter snakes caught were "pets," and Josephine, the big dalmatian, would sleep with them for protection and warmth. Given minimal instruction on how to find a stream, know the north side of a tree even when cloudy, no one ever got lost. I am suggesting that what I am calling *flatness* is very unlike Heidegger's "standing reserve" or "Gestell" which is a reductive notion colored by the late modern efficiencies of the industrial technologies. Heidegger would have a problem thinking a cell phone could be "ready to hand," but that is what it is to today's users. But, unlike the at-best local or regional reach of a hammer, the cell phone reaches a global world.

Jump now to the lifeworld of the popular or ubiquitous technologies just noted: I would argue that this world is also "flat" insofar as, for example, the pluriculture which is mediated via TV, cell phone, cinema, and even the musics of the earbud technologies are now a global interlinkage. But, the texture of this flatness is not that of a "Global Village." The near-distance of cyberspace time is not that intimate. It is malleable, plastic, multistable, yes, but it is not face-to-face (see *Listening and Voice).* Combined, the networking is also the world of *social media* such as the blogs, chats, Facebook, and countless other connectors. It is a wide world, far more diverse than village culture, pluricultural. Some argue that however rich, such interconnection is disembodied—I deny this, but it is only a partial embodiment, a mediated embodiment. And just as noted with AV/OA, electronic interconnection is not a whole body connection, but it is a connection. What follows is somewhat speculative. In such quasi-situations could it be that like the echo-locating blind bicycle rider, one may *amplify* other sensitivities either to compensate, or in many of the social media situations, overamplify experiential feelings? E-mail romances (which sometimes turn sour or dangerous) or the recruitment of terrorists for very different religio-cultural tasks, are not infrequent results of contemporary interconnectedness. But, if I look at my own life post-Internet, like many academics one can meet first online, only later to meet face-to-face. Our lives alternate between the "worlds" of electronic connections and face-to-face connections. Or, take the slow process by which contemporary teenagers become critically aware of the possible consequences of sexting, drinking, or other displays which do not get erased from the Internet. Politicians today are often tripped up by just such social media events. It is clear that social media do not determine such outcomes or events, but it is easily possible to enter texts, pictures, or other AV/OA data on the Internet, and once there it is available to the world and for the most part permanent. Europe has become concerned with the problem of privacy and erase ability and is moving toward versions on erasure legislation. Most American companies involved with the Internet oppose such legislation. It is part of the Big Data/privacy and NSA surveillance controversy of the present. I shall not enter that issue here since it calls for a book-length discussion.

PART TWO: TWENTY-FIRST CENTURY AUDIOVISUAL/ OPTO-ACOUSTIC TECHNOLOGIES

I want here to return to the more special function contemporary AV/OA technologies, some of which are in developmental stages. In this final update context, I want to emphasize how much contemporary technology focuses upon and utilizes *micro-processes*. Many philosophers of science have commented upon how our very largest and most complex instruments, such as the planned but cancelled Supercollider in Texas, and today the CERN Large Hadron Collider, are designed to image and manipulate the subatomic. In 2012, thousands of scientists used the CERN collider to produce the "God particle" or the Higgs Boson; in 2015 a likewise massive collection of investigators have produced the *penta-quark* which is a subatomic particle associated with protons—the biggest instruments for the smallest entities.

Whether the instruments themselves are large or not, much of AV/OA specialized technologies or instruments are focused upon micro-phenomena. Return first to several of the previously discussed acoustic technics: I include here the bugging devices of the Cold War in which a laser reader picks up the micro-frequency waves of speaker diaphragm substitutes which vibrate due to voices inside a building. But similarly, also as noted, CD, earbud, music devices use variants of the same technics—a laser "reads" a series of bumps and flats on a disc. This system is opto-acoustic in the sense that the reading technology is not physical, but optical although the output is acoustic, the listener hears the music. It is a quantum step into micro-phenomena compared to, say the Edison phone or cylinder recordings which relied upon some form of physical reader of the bumps and groves of the wave. Technically, a digital process is a nanoprocess and each CD *pit* (a zero in the binary digital code) and *land* (a one in the same code) is but two nanometers in size (two millionth of a nanometer which is one millionth of a meter). Optoacoustic technics are today, virtually the standard process for such instruments.

Return to medical diagnoses for cancers. I emphasized the sonification version of cancer detection whereby the data (the code of numbers) is turned to sound images rather than visual images and thus the distinctions between healthy and cancerous cells can be heard. In a sense, this is the simplest albeit possibly also the least expensive and effective acoustic means of diagnosis. But there are other ways in which optoacoustic technics can work as well. This technology, hybrid in form, is still in its infancy. The scientific claim is that such imaging ideally combines the accuracy of spectroscopy with the depth resolution of ultrasound. Spectroscopy, early was the key to chemical signatures of the stars. In today's mass spectroscopy the analysis of ions can identify from exactly where a volcano produced a particular obsidian. Likewise, auditory surface penetration or depth imaging lies within ultrasound imaging. Here are some examples: laser optoacoustics, because the focus can be very small, is aimed at diagnosing very tiny breast cancers, too small to be detected by mammography (1999); photoacoustic imaging (2007), possible to resolve tumors as small as 200–1000 nanometers is aimed at diagnosing prostate cancer. Multifrequency microwave thermal imaging, combined with acoustic imaging (2007) is another technics. Finally, pancreatic cancer, very hard to detect early due to cells circulating in the bloodstream is also beginning to be imaged by optoacoustic enhanced processes. All of these hybrid technics rely upon microprocesses not developed until the twenty-first century. Most of the technics just noted are not yet widely available, but are cutting edge processes which are extant and being developed.

NOTES

1. Evan Selinger, "Does Microcredit Empower? Reflections on the Grameen Bank Debate," *Human Studies*, 31, 2008, pp. 27–41.

2. Galit Wellner, *A Postphenomenological Inquiry of Cell Phones: Genealogies, Meanings, and Becoming* (Lexington Books, 2015).

3. Peter Pollack, *The Picture History of Photography* (New York: Harry N. Abrams, Inc. Publishers, 1977) pp. 65–68.

4. Don Ihde, *Listening and Voice: Phenomenologies of Sound, 2nd edition* (Albany: SUNY Press, 2007), pp. 161–166.

5. Stacey Irwin, *Digital Media: Human-Technology Connection* (forthcoming Lexington Books, 2015).

Are We Posthuman?

That we are postmodern, I do not doubt. In standard historical terms, "modern" is both ambiguous and chronologically variable. Early modern science, in the standard view, begins with the seventeenth century. It turns mathematical for an interpretive framework, and experimental for investigation and verification, both processes exemplified by Galileo with his mathematical language of the universe and his inclined planes and swinging chandeliers in Pisa. Contemporary science is such, to use only one variable—instruments—that no one would return to the limits and practices of Galileo. Modern art, by the same standard history, comes much later, with Dadaism, Cubism, and other twentieth century art styles. I have done quite a bit of art history study on portraits and it is clear to me that artists are no more likely to return to Renaissance style portraiture today than scientists are to return to 30X Galilean handcrafted telescopes. Top selling portraits from Andy Warhol, Lucien Freud, Gerhardt Richter, and others simply are unlike those of the Renaissance. As should be seen, I am using "postmodern" to designate a style which differs from that of modernism, not an ideology. In my sense we are postmodern if our practices and sensibilities are such that we are unlikely to return to modern practices and sensibilities. The more ideological version of postmodern, most loudly claimed by Jean Francois Lyotard, is also twentieth century. His *The Postmodern Condition* (1979 French/1984 English) became a sort of "bible" of theoretical postmodernism, with the death of a master narrative and the emergence of multiperspectivalism. Much of my own work has been in science-technology studies which have surely undergone a revolution since the mid twentieth century, with philosophy of technology showing how science is embodied in its technologies—technoscience. Neither science nor technology looks like it did even a few centuries ago. In my sense, the "postmodern" is a change in practices and style which is no longer "modern."

Are we, however, *posthuman*? As my thesis has unfolded here, it has become increasingly obvious since the nineteenth century that there are clearly recognizable human horizons, boundaries, regarding our bodily experience of the world. Even our best sensory openness to light and sound, remains limited and is exceeded by many of our animal relatives whose sight and hearing is open to infrared and ultraviolet, to infrasound and ultrasound, and in other cases have perceptual awareness of the world—thermal, magnetic—which is either so minimal with us humans, or not present at all, that we are more aware today of human limitations. On the other hand, through technologies, instruments, human experience has clearly been expanded through technological mediation. Our world is starkly different from what it appeared to be even a century ago. In astronomy, black holes, pulsars, neutron stars, multiple galaxies in multiple shapes, were neither theorized nor perceived in earlier times. Much, too, has disappeared, such as phlogiston, aether, Democritean atoms, and more recently event horizons. In this disparity between recognizing our contingent and finite capacities and our amplification of powers through technologies, does a fantasy regarding making ourselves superhuman thrive?

Nor is the instrumental mediation only perceptual. Our instruments *intervene* as Ian Hacking noted,[1] and today such devices as laser tweezers and related micro-tools can "trap" atoms and depict single photons (with computer and laser assisted femtophotography). Our reach—with the Pluto flyby July 2015, from billions of miles; drone warfare with autonomous airplanes piloted over Afghanistan from the American Southwest; the Mars Explorer; reaches from the intimacy of laparoscopic surgery to billions of miles of flyby photography. So, while it might be taken that we today recognize our bodily-perceptual limits, our technological reach has been vastly extended.

One philosophical response to this new twenty-first century situation has been the emergence of a series of "beyond" human—I shall call them—*technofantasies.* Minimally these run through virtual reality mini-worlds, to reality enhancement experiments, to more global notions such as the posthuman and transhuman. These ideas and the associated technologies, form a continuum which I want to briefly explore in this epilogue. I remain both critical and skeptical of technofantasy, in part because after now more than forty years of investigation and research into both science and technology studies, history, and analysis, I have come to recognize that particularly at early stages of technology development, there is inevitably a first "utopian" and "dystopian" set of fantasies which occur.

I will here begin, as I frequently do, with a personal relationship to what might be called "becoming a cyborg." It was Donna Haraway who made *cyborgs* as human-animal-technology hybrids popular.[2] In my own case, I sometimes call myself a "Cyborg Emeritus." In a series of articles on the theme "Aging: I don't want to be a Cyborg,"[3] I reflected upon how slowly, then more rapidly, I have become cyborgean. Beginning long ago, but with a new one this last year, I had teeth topped with crowns; then, once over seventy, an arterial stent, later still, after mitral valve repair (thus no animal parts yet) and bypass surgery, a plastic valve seat, then knee replacements in stainless chrome steel and plastic, and most recently a paraesophageal hernia repair via laparoscopy. X-rays show how many staples are scattered through my body. So, while I might not have wanted to be a cyborg, what in each of these procedures eventuated, was that it is better to be cyborg than being impaired or dead! But as I learned from etymological histories, cyborg results are not "bionic" which term usually associates with a notion that the replacement technology is "better" than the previous biology. That remains a technofantasy. This terminology reflects what I am calling a technofantasy associated with some vectors of the posthuman.

My strategy here will be to look at the continuum of technologies which get associated with posthuman, transhuman, enhanced, and virtual reality movements and do what I take to be a postphenomenological critique. I begin with what could be called detachable prostheses:

- Limb prostheses go back at least to ancient Egyptian times (foot replacements found in archaeological remains), and gradually change (peg legs for pirates, hooks for hands) until today's hi-tech leg replacements, hand and arm replacements. Vivian Sobchack and Helena De Preester have excellent discussions of these.[4]
- Sensory prostheses, such as eyeglasses, contacts, hearing aids, in today's hi-tech versions are far better than those of even decades ago (see chapter 7). And one could continue with improved detachable versions such as lenses used by diamond cutters, dentists, ophthalmologists in optics. Or acoustically, the specialized versions of stethoscope listening devices used by mechanics. These latter are simple "enhancement" technologies which image beyond our usual sight and sound capacities. Enhancement prostheses are usually detachable for good reason, if one had a color filtered magnification permanent eye change, it would be hard to navigate even in familiar surroundings.
- I shall as an aside include a vast group of what could be taken as tactile-kinesthetic "protheses" although we do not usually think of such technologies as prosthetic. I refer to machines which dig, probe, lift, controlled by embodied humans. Excavators, jack hammers, cranes, and a plethora of like machines multiply "muscle power" which magnifies human capacity manyfold.

This list is at best suggestive. These are for the most part "analogue" technologies in that most, as in my own surgery examples, are designed to restore capacities, although the specialized examples change or magnify capacities. The technofantasy dimension, however, can be found in the histories of most of these technologies. The folk legend and songs associated with "John Henry" come to mind. Here human, John Henry, is pitted against machine, in building a railway. In the legend John Henry wins, but dies in the process—yet, who today would want to return to the hard labor of hand building a railway? I, for one, enjoyed seeing the power and the skill of the gigantic jack hammer excavator which was used to smash a bluestone boulder which sat just in the middle of where my garage-studio was being built. With a couple of blows in minutes, the rock exploded into hundreds of fragments. Had this been done with nineteenth century Vermonters with sledge hammers, it would have taken a week to accomplish. Part of my point is that what, on first arrival, technologies once embedded and embodied, no longer fit their early "hyped" fantasy modes. They become taken for granted and lose their fantasy mystique.

Next on the continuum with external and removable prostheses, are permanent implant technologies, most of which are new and many still experimental:

- For acoustics, the *cochlear implant* which is a digital and programmed hearing device very different from hearing aids. It has a programmed voice production which is directly wired into the user's ear physiology and is implanted in the user's skull. Technically, it is basically a mini-computer which receives sounds, is speech selective, and transmits processed sounds to the inner ear. These devices are expensive, hard to learn to use, and became practical only by the late twentieth century (primary development, 1976–1997). Roughly half a million users have implants worldwide. A very descriptive autobiographical account was given in *Rebuilt: My Journey Back to the Hearing World* (2006) by Michael Chorost, and an in depth postphenomenological analysis has been done by Kirk Besmer.[5]

- Older and more common are cardiac pacemakers. These, too, are twentieth century devices which initially used radio wave technology to provide a regular single speed pace signal to the heart to control beat, and more recently to variable speed controls. The first (large and external) pacemaker was invented by John Hopps in 1950, later implantable pacemakers by Wilson Greatbach were used. I do not have a pacemaker but my own stent was invented in the same period, but is a passive wire mesh device which simply keeps an artery open.

- Perhaps today's most hyped implants are those associated with deep brain implants, most designed to combat pathologies such as epileptic seizures, Parkinson's attacks, and most recently tinnitus ear ringing episodes. Successes are ambiguous and experimentation continues.

- Related technologies relate to restoring or providing lost sensory capacities—especially for the blind. Various patterns of stimulation delivered to one's retina, or even tongue, sometimes provide very rough visual-analogue imagery. But as per my cyborg observations, these are at best approximations.

- Finally, I need to mention the "bionic" brain wiring technologies such as those which mediate brain control presumably through thought of such things as moving cursors on screens to in turn manipulate limbs, a sort of executive consciousness technology which extends nerve responses through distance technologies. The hope is to develop more controllable and sensory feedback prostheses for lost limbs like hands.

Before turning to the posthuman and beyond issues, I wish to issue a critical warning similar to that which I used in the technoscience research seminar and in advice to writers. The caution is to always ask "who says this?," "who benefits?," and what is the realistic or empirical outcome? In today's interconnected world, each new invention is early hyped. First, there is what could be called *manufacturer hype.* Each maker extols the breakthrough, the benefits of the product. But early hype does not bear caution. Take for example the production of joint implant technologies. These, often designed by surgeons but manufactured by engineering firms, does variants upon the various parts of the device. But as with auto industry recalls, recent metal-on-metal joint implants have had to be "re-called" due to metal filings getting into the user's body. But here a "re-call" entails another surgery!

The second hype layer is *media hype.* Science news is big business. The National Science Foundation budget for education (which includes advertising through releases and documentaries) is larger than the entire National Endowment for the Humanities budget. Each new invention or breakthrough is publicized, often with hype. Media hype runs through everything from new telescopes to pharmaceuticals. It would not be hard to produce a "top ten" list as popular music radio used to do. As one example, today the development of a self-driving or autonomous car would surely be one item on the list. A critical followup often does not occur, and lacking postintroduction critique it is easy to assume all is going well. But as the auto industry repeatedly illustrates there are often some very large consequences which could be a multi-million car recall. (And these do occur as with exploding gas tanks or faulty ignition switch problems with Ford and General Motors in recent decades.)

Now the turn to the question of this epilogue: Are we posthuman, transhuman, or beyond? In this case I will invert the end of the continuum of technofantasies and begin first with the most extreme and then work down to the more modest of these. *Transhumanism* has a complex history and for the sake of briefness I shall not trace any of these movements deeply. There are strands of thought common to the leading transhumanists with early inspiration coming from an almost religious futurology, much from England (early Julian Huxley, J. B. S. Haldane, Arthur Clark) with much drawn from science fiction thinking. Early artificial intelligence, robotics, and above all a utopian set of hopes for rapidly changing technologies. Has Moravec, a futurologist AI robotics thinker proclaimed the hope for downloading the human mind into a computer (the computer presumably being eternal); Ray Kurxweill predicts a coming "Singularity" due to ever faster technological development, including the hope for technologically aided eternal life; Nick Boston proclaims like improvements to the human body, longevity, and mental enhancement. What emerges in these contexts is, on the one side, the recognition of human limitations, but on the other a very utopian, science fiction belief in the inevitability of technological perfection and speed of development, and for virtually side effect free results. I, myself, have often been at conferences with some of these principals. What seems to me to be the case, is that with the exception of a technological twist, I see nothing new in transhumanism. Instead, its desire for everlasting life, for a stronger, smarter, better body seems to me to reflect human desire fantasies as old as antiquity. The literal reductionistic physicalism (downloading a mind into a computer), more fictional than scientific, but now concentrated technofantasy form, simply perpetuates an ancient human dislike of contingency and finitude. Granted, the means for fulfilling the fantasy are no longer magical or animal based, but in the hoped for technologically inevitable perfectibility of the future.

Posthumanism, less technoutopian than transhumanism, comes in several varieties. Its postmodern form rejects what could be called "Enlightenment Humanism," the interpretation of the human associated with the eighteenth century and early modern thought. Here the "human" is highly autonomous, rigorously private, individualistic, and endowed with eternal "rights." Narrow posthumanism is a recognition that this description of the human is flawed—humans are more social, interconnected, in a Merleau-Pontean sense "outside themselves in the world" rather than locked inside a subjectivity. Here I find sympathy. However, in a technofantasy mode, posthumanism can also hope for some versions of "improvement" through bodily, perhaps biological improvements. Critically, what worries me is that the temptation for post-phenomenologists is that too much acceptance of manufacturer-media hype can be swallowed and thus a slope closer to the transhuman appears. While I advise that we have no fears about "turning Hal off" from *2001: A Space Odyssey*, neither is the prospect of generations of millennium-long lived humans encouraging. Perhaps we need a new Jonathan Swift to depict an ever expanding population, all living to be millennial lived, until the surface of the earth is covered. Both utopianism and dystopianism are technofantasy variants. A hard look at a postphenomenological "realism" and a critical but empirical analysis is the better option.

In 1997, Hans Achterhuis, one of the Netherlands' leading philosophers of technology published a book which in English would have the title, *From Steam Engine to Cyborg: Thinking Technology in the New World*. It was unique; he recruited his philosophy of technology colleagues to each do a chapter on six American philosopher's of technology—my generation—to contrast them with earlier European philosophers on technology. We took to translate this into *American Philosophy of Technology: The Empirical Turn* (2001). Peter Paul Verbeek, probably today's most prominent and visible postphenomenologist, was assigned "Don Ihde: The Technological Lifeworld." It is an excellent survey and analysis of my turn to postphenomenological analysis and Verbeek looks in detail into the ways Martin Heidegger and I deeply diverge on both phenomenology and technology. In one summary sentence Verbeek claims, "Instead of questioning 'backwards' [Ihde] questions 'forwards,' that is, instead of reducing technological artifacts to the technological for of world-disclosure that makes them possible, he takes what form of world-disclosure is *made possible by* technological artifacts." [6] But that shift is made possible be a much earlier discovery. *Experimental Phenomenology* (first edition, 1977) was an attempt on my part to actually practice phenomenological variations, in that case on what were then called "ambiguous drawings." But as I applied what I took to be Husserl, I discovered not an "essence," but *multistabilities.* Later, turning to much more material phenomena, and particularly to technologies, it became obvious that multistability was a deep characteristic of many phenomena, and particularly of technologies. Thus, as I began to fashion a "postphenomenology" multistability became a forefront feature—this is also why Verbeek rightly sees my emphasis upon "forward" perspectives which forefront multiple possibilities (Acheulean axe to cell phones). What he does not note is that this movement is also one which incorporates American pragmatism with its anti-essentialism and anti-foundationalism.

So, with regard to the posthuman, am I, cyborg emeritus, posthuman? So far, I think not. Rather the now literally incarnated technologies—which in no way have made me bionic or superhuman—have allowed me to continue a long living life, to engage in social life, and to be part of the pluricultural conversation with its alternatives between technologically mediated and face-to-face encounters. Even what could be considered my perceptual enhancements, if I use night vision goggles, my backyard telescope, or play with virtual and enhanced reality technologies, all these remain detachable. To listen to the new sounds, but also find peaceful silences, remains human and embodied.

NOTES

1. Ian Hacking, *Representing and Intervening* (Cambridge: Cambridge University Press, 1983).

2. Donna Haraway, *Simians, Cyborgs, and Women: The Reinvention of Nature* (New York: Routledge Publishers, 1991).

3. Don Ihde, "Aging: I don't want to be a Cyborg." *Ironic Technics* (USA & UK: Automatic Press, 2008), pp. 31–42.

4. Helena De Preester, "Technology and the Body: the (Im)possibility of (Re)embodiment," *Foundations of Science*, 16 (2), 2007, pp. 119–137. Vivian Sobchack,

Carnal Thoughts: Embidiment and the Moving Image Culture (Los Angles: University of California Press, 2004).

5. Kirk Besmer "Embodiment a Translation Technology," *Techne: Research in Philosophy and Technology,* Vol. 16 (2), fall, 2012, pp. 296–316.

6. Peter Paul Verbeek, "Đon Ihde: The Technological Lifeworld," in Hans Achterhuis, *American Philosophy of Technology: The Empirical Turn,* (Bloomington: Indiana University Press, 2001), pp. 12–13.

Technoscience and the Twenty-First Century [1]

Many people have begun to notice that the type and style of technologies being produced in very recent years differ from those, even of the last century. The mega-technologies of the nineteenth and twentieth centuries were "Industrial" technologies, correctly characterized as gigantic, mechanical, and industrial by Martin Heidegger. The new technologies, recognized by the EU and others are more dominantly "electronic"—not even electric which itself still falls under the industrial.

Less noticed is the drive to the micro-levels of material manipulation within new technoscience—look at several of these:

- Nanotechnologies. These technologies, still largely developmental, are micro-machines composed of only a few atoms or molecules, film surfaces, tubes, and the like which still are surrounded by technohype with dreams of what they can do, but insofar as examples have been created are microscopic or smaller.
- The biotechnologies. Here everything from the transplanting of bio-luminescent genes into various animals for "green" rabbits, hogs, and other creatures, to genes which resist freezing, transferred from flounders to tomatoes (Haraway's example), to GM seeds with vitamin A or pest resistance are again microscopic level technologies.
- IT and ICT technologies. The Internet and all its cousins operate electronically. Miniaturization is a part of and a desiratum for these technologies as well. The smaller the processors the better. One forgets the mega computers which once took up whole MIT basements. The dream here is of atomic level computers.
- Imaging technologies. These are the technologies which I shall concentrate upon, technologies which again with microscopic material manipulation, can produce data and images (both visual and acoustic) of both distant and nearby phenomena ad infinitum. It has been my contention that a *second scientific revolution* has been occurring with imaging technologies from the late twentieth into the twenty-first centuries
- However, before turning to this forward-looking phenomenon, I wish to enter a caveat regarding the industrial technologies. These still remain in place and new ones are even being produced. It seems that technological ages do not simply replace one another,

but continue alongside one another—at least for a time. Technological ages do disappear—the stone age with its cabinet of tools and
weapons, is gone except for archaeological speculation and museums, and yes, I have my own favorite Acheulean hand axe which
weighs down papers in my Vermont study! Steam locomotives
only remain in a few underdeveloped countries and as tourist
trains. But vast industrial nets and lines stretch across ocean spans
to factory-capture fish; mega-scissors machines snip off large trees
and deforest vast rainforests; and factory farms produce everything
from chickens to salmon. In short, industrial technologies even if
laser in a process not unlike its role in "reading" a CD disk, "reads"
window vibrations and through the technology reproduces the
conversation. Here, again, there is a micro-process which detects
waves and translates them into intelligible voices. One might note
that this surveillance setup is simply a more complex variation on
similar processes in CD players. This micro-vibration reading is a
deep part of contemporary imaging and today even concrete col-
umns can be "read" for micro-vibrations.

These capacities, one can see, very much depend upon the electronic and
micro-level phenomena of second revolution technologies. And they produce very different problems for our lifeworld. From my perspective, I
would say the set of most serious problems related to the industrial technologies ended up being *environmental problems*. The homogenic production of global pollutants, from gases, particles, and other effluents, to
ground and oceanic degradation were the spawn of mega-technologies,
problems still with us and still unrestrained.

The problems of the upcoming e-technologies seem to me to be more
socio-political. Social media, big data collections and uses, screens as attention holders, surveillance, and robotics end up relating to socio-political problems which now dominate the impact from human-technology
relations. I will briefly look at a few examples.

BIG DATA

This is the term which loosely applies to the vast sea of information
produced by the various IT. ICT and other electronic media from the
Internet to phones to e-mail and the like. The current large scandal comes
from the Snowden revelations concerning US NSA collecting of this data
from virtually the entire world. Earlier WikiLeaks also uncovered this
vast enterprise on the diplomatic level. If it may be e-transmitted it may
be captured and stored. As with all technologies, one could foil this process by unplugging. At the moment this scandal seems centered upon
governments—but businesses, political groups, and multiple others also

desire big data to mine for social-political and economic reasons. Note how different this set of problems is from CO_2 emission problems or atmospheric lead or mercury poisonings. I can't resist pointing out that the single largest source of mercury in the atmosphere today comes from the practices of artisanal gold mining, not factory processes!

SURVEILLANCE

Although big data overlaps with surveillance, visual surveillance is more image related. From near omnipresence of video surveillance—in my apartment building to the supermarket—to the highest concentrations in the UK, the populace is "watched." Enter also robotics which increasingly are embodied in small, even insect sized imaging robots. The military has model airplane sized optical robots and is developing bumblebee and even fly sized ones. There are systems for parents to observe babysitters and empty houses. And, of course, voyeurism has become much easier. But then, so has geriatric oversight. On a more positive note, camera capture devices, crittercams, and the like have vastly expanded what we know about wildlife behavior.

SCREENS

I use this generic term to cover a vast spectrum of contemporary imaging technologies. The US collegian is known to spend, on average, twelve hours per day either watching some screen (or more than one) or plugged into a music device. From cinema to television to laptops to iPads to iPhones, screens concentrate focal attention—here, too, the social media. Positive regarding information and entertainment; negative regarding addictions and distractions, screen watching captures a lot of twenty-first century human time. It transforms social time. The demographics also show that while flat TV and cinema screens are larger, by far the most watching is on small—from laptops to phones—screens.

CELL PHONES

Although cell phones are simply one example of the above, I list them separately because according to sociologists they are probably the single most ubiquitous technology in today's world. The claim is that 95 percent of the world's population has some access to cell phones which in turn can reach globally with connections. One of my favorite images is of a

Masai herdsman, spear in one hand, cell phone in the other. This is a "leapfrog" technology which places easily in underdeveloped areas since its infrastructure is not hard wired systems but simple cell towers for transmission. It is also electronic, miniature, and multipurpose.

NOTE

1. This is an abbreviated version of my Lisbon panel presentation, 2013. The panel was composed of six charter or founder members of the Society for Philosophy and Technology. Our task was to project technology in the twenty-first century.

Bibliographic Essay

Acoustic Technics is my first book with our new series on "Postphenomenology and the Philosophy of Technology." I am rightfully proud of this publication event. This series with Lexington Books marks a mature culmination of events which go back in both my own philosophical history to at least 1984 and for what became today's postphenomenological movement, first associated with the Technoscience Research Group at Stony Brook University, officially begun in 1998 through the end of my Stony Brook career with retirement in 2012. Here I want to highlight some crucial formative points and emphasize their bibliographic outcomes.

As noted elsewhere, in lectures at Goteborg University in 1984, I argued for a "Non-Foundational Phenomenology" inspired by Richard Rorty's nonfoundationalism of the time. By 1993, realizing this term was too clumsy, I opted for *Postphenomenology: Essays in the Postmodern Context* and the term stuck. By 1994, a trickle of visiting scholars began to congregate at Stony Brook from which the Technoscience Research Group was formed and a permanent Technoscience Research Seminar became part of our PhD program by 1998. This group concentrated upon the philosophies of science and technology and cultural and STS studies interdisciplinary work, including our famous "roasts" of the major figures of these disciplines. Almost at the same time, Peter Paul Verbeek was doing his final doctoral work and a substantial part of his early publication concentrated upon the shift from classical to postphenomenology. His career, if we use today's term, "went viral," and today I would think him to be Europe's most visible postphenomenologist. His group at Twente is dynamic and has many MA and PhD students. Both his English books, *What Things Do* (2005) and *Moralizing Technology* (2011) are internationally read. Another center of activity today is in Denmark, at both Aarhus University and the Danish Pedagogical University with Cathrine Hasse and Finn Olesen as principals.

Over the years at Stony Brook dozens of doctorates were completed; scores of visiting scholars and postdoc and faculty visitors participated, and by 2007 on, we began, with collaboration with Europeans and Asians, to participate in the "Postphenomenology Research Panels" which still take place at STS styled conferences (SPT, 4/S, SPHS, and elsewhere). From these, many papers were published and various journals such as *Human Studies, Foundations of Science, Techne* included both individual articles and special issues. Our Lexington series now includes many of these papers in volumes out and forthcoming. The Robert Rosenberger, Peter Paul Verbeek *Postphenomenological Investigations* (2015) is out and the Jan Friis, Robert Crease *Technoscience and Postphenomenology: The Manhattan Papers* (2015).

Here, however, I want to concentrate upon research most relevant to *Acoustic Technics*. Postphenomology has adapted a style similar to cultural and STS studies which favors "case studies," or thematic work on particular technologies and issues. *Acoustic Technics* is deliberately a very contemporary followup of *Listening and Voice* (first edition, 1976, second edition, 2007) and forefronts acoustic technologies. And although I cite many of the related works in footnotes, I want to summarize those closest here in this bibliographic essay:

- Ultrasound Technologies, especially sonograms: Peter Paul Verbeek, in both books noted, and, "Obstetric Ultrasound and the Technological Mediation of Morality: A Postphenomrnological Analysis," *Human Studies* 31: 11–26.
- Cochlear Implants: "Embodying a Translation Technology: The Cochlear Implant and Cyborg Intentionality," (Besmer) *Techne: Research in Philosophy and Technology* 16 (3): 296–316.
- Prostheses: Helena De Preester, "Technology and the Body: The (Im)possibilities of Re-embodiment," *Foundations of Science* 16: 119–137.
- Cell Phones: Galit Wellner, *A Postphenomenological Inquiry of Cell Phones: Genealogies, Meanings, and Becoming* (forthcoming, this series) And, Robert Rosenberger "Embodied Technology and the Problem of Using the Phone while Driving," *Phenomenology and the Cognitive Sciences* 11(1): 79–94.
- Pap Smear reading: Anette Forss, "Cells and the (Imaginary) Patient: The Multistable Practioner-Cell Interface in the Cytology Laboratory," *Medicine, Health Care and Philosophy* 15: 295–308.
- Cultural Perception: Cathrine Hasse, "Postphenomenology: Learning Cultural Perception in Science," *Human Studies* 31 (1): 43–61.
- Earbud Music Technologies: Stacey Irwin, *Digital Media: Human-Technology Connection* (forthcoming, Lexington Books).

Others appear in the footnotes and text, but I want to conclude by pointing out two books which include a number of essays directed toward my own analysis of postphenomenology: Evan Selinger, *Postphenomenology: A Critical Companion to Ihde* (2006) and the J. K. B. O. Friis and Robert Crease, *Technoscience and Postphenomenology: The Manhattan Papers* (2015).

Index

Lightning Source UK Ltd.
Milton Keynes UK
UKOW01n2344110117

291832UK00010B/135/P

9 781498 519236